Introduction to Cell and Tissue Culture
Theory and Technique

INTRODUCTORY CELL AND MOLECULAR BIOLOGY TECHNIQUES

SERIES EDITOR: Bonnie S. Dunbar, Baylor College of Medicine, Houston, Texas

INTRODUCTION TO CELL AND TISSUE CULTURE: Theory and Technique
Jennie P. Mather and Penelope E. Roberts

Introduction to Cell and Tissue Culture

Theory and Technique

**Jennie P. Mather and
Penelope E. Roberts**

*Genentech Inc.
South San Francisco, California*

PLENUM PRESS • NEW YORK AND LONDON

Library of Congress Cataloging-in-Publication Data

Mather, Jennie P., 1948-
    Introduction to cell and tissue culture : theory and technique /
Jennie P. Mather and Penelope E. Roberts.
        p.   cm. -- (Introductory cell and molecular biology
techniques)
    Includes bibliographical references and index.
    ISBN 0-306-45859-4
    1. Cell culture.   2. Tissue culture.   I. Roberts, Penelope E.
II. Title.   III. Series.
    [DNLM: 1. Tissue Culture--methods.   2. Microbiological Techniques.
QS 525M427i 1998]
QH585.2.M38   1998
571.5'38--dc21
DNLM/DLC
for Library of Congress   98-27597
                                                                   CIP

ISBN 0-306-45859-4

©1998 Plenum Press, New York
A Division of Plenum Publishing Corporation
233 Spring Street, New York, N.Y. 10013

http://www.plenum.com

10 9 8 7 6 5 4 3 2 1

Printed in the United States of America

To the memory of Dr. Izumi Hayashi,
whose life was as elegant as her experiments

Foreword

It is a pleasure to contribute the foreword to *Introduction to Cell and Tissue Culture: Theory and Techniques* by Mather and Roberts. Despite the occasional appearance of thoughtful works devoted to elementary or advanced cell culture methodology, a place remains for a comprehensive and definitive volume that can be used to advantage by both the novice and the expert in the field. In this book, Mather and Roberts present the relevant methodology within a conceptual framework of cell biology, genetics, nutrition, endocrinology, and physiology that renders technical cell culture information in a comprehensive, logical format. This allows topics to be presented with an emphasis on troubleshooting problems from a basis of understanding the underlying theory.

The material is presented in a way that is adaptable to student use in formal courses; it also should be functional when used on a daily basis by professional cell culturists in academia and industry. The volume includes references to relevant Internet sites and other useful sources of information. In addition to the fundamentals, attention is also given to modern applications and approaches to cell culture derivation, medium formulation, culture scale-up, and biotechnology, presented by scientists who are pioneers in these areas. With this volume, it should be possible to establish and maintain a cell culture laboratory devoted to any of the many disciplines to which cell culture methodology is applicable.

Dr. David Barnes
Department of Biochemistry and Biophysics
Oregon State University

Acknowledgments

We would like to thank Dr. David Phillips for all we have learned about looking at cells during many years of collaboration. Thanks also to Dr. Phillips for providing the scanning and transmission electron micrographs used throughout the book. Our thanks to Dr. David Barnes for many interesting discussions on the nature of cells, from worms to man, over many years. We would also like to thank Dr. Barnes, Dr. Monique LeFleur, Amy McMurtry, and Patricia Kaminsky for their careful reading of draft versions of the volume and their suggestions for corrections and clarifications. We would also like to thank Aldona Kallok for helping in many ways with the preparation of the manuscript.

We would also especially like to thank Alicia Byer, Dr. Lin-Zhi Zhuang, Dr. Virgilio Perez-Infante, Mary Tsao, Robert Shawley, Diana Stocks, Dr. Margaret Roy, Dr. Yossi Orly, Dr. Teresa Woodruff, Dr. Alison Moore, Dr. Rong-hao Li, Dr. Jean-Philippe Stephan, Dr. Vidya Sundaresan, Terri Restivo, Marcel Zocher, Kathy King, Glynis McCray, and the other past and present members of our laboratory. It is impossible to overestimate the contributions of these friends and colleagues who have, in the course of their work and studies in the Mather Laboratories over the years, added greatly to our knowledge and the fun of cell culture. Finally, we would like to thank Dr. Gordon Sato, who introduced us to the joy of cell culture and the infinite variety of interesting things to do with cells.

A note on the figures: The graphs and tables presented throughout the book are drawn from actual experimental data generated in the Mather Laboratories over the last 20 years. We have chosen those experiments that best illustrate the point being discussed in the text and have not necessarily provided all the experimental details for each figure.

We would also like to thank the following vendors for their help in discussions of their equipment and, where noted, in providing photographs or data for the figures and tables: James Quach, Instrument Services, Genentech, BRL Life Technologies, Corning Corporation, Falcon (Becton Dickinson), The Baker Company, Mike Alden of Coulter Electronics, E. Braun Biotech International, The Edge Scientific Instrument Co., Altair Gases, Sara Ferrer and Technical Instrument Company, and Brent Kolhede of Lab Equipment Company.

Contents

Introduction to Cell
and Tissue Culture
Theory and Technique

CHAPTER 1

Introduction

THE HISTORY OF TISSUE AND ORGAN CULTURE

Tissue culture as a technique was first used almost 100 years ago to elucidate some of the most basic questions in developmental biology. Ross Harrison at the Rockefeller Institute, in an attempt to observe living, developing nerve fibers, cultured frog embryo tissues in plasma clots for 1 to 4 weeks (Harrison, 1907). He was able to observe the development and outgrowth of nerve fibers in these cultures. In 1912, Alexis Carrel, also at the Rockefeller Institute, attempted to improve the state of the art of animal cell culture with experiments on the culture of chick embryo tissue:

> The purpose of the experiments . . . was to determine the conditions under which the active life of a tissue outside the organism could be prolonged indefinitely. It might be supposed that senility and death of cultures, instead of being necessary, resulted merely from preventable occurrences; such as accumulation of catabolic substances and exhaustion of the medium. . . . It is even conceivable that the length of life of a tissue outside the organism could greatly exceed its normal duration in the body. (Carrel, 1912, p. 9)

Carrel succeeded in expanding the possibilities of cell culture by keeping fragments of chick embryo heart alive and beating into the third month of culture and growing chick embryo connective tissue for over 3 months. Using apparatus such as that shown in Fig. 1.1, Carrel reported growing chick embryo tissue for many years *in vitro,* and thus helped convince the scientific community that *in vitro* cultures were useful experimental systems.

The next important advance in the conceptualization and technology of cell culture was the demonstration by Katherine Sanford and co-workers (1948) that single cells could be grown in culture. This, along with Harry Eagle's (1955) demonstration that the complex tissue extracts, clots, and so forth previously used to grow cells could be replaced by ". . . an arbitrary mixture of amino acids, vitamins, co-factors, carbohydrates, and salts, supplemented with a small amount of serum protein . . . " (p. 50) opened up a new area of cell culture. A vast range of manipulations that had not been possible previously could now be per-

Figure 1.1. A photograph of the tissue culture apparatus such as that used at the turn of the century.

formed with cells, including production of genetically altered cell lines through mutagenesis and cloning, direct comparison of cells from normal and transformed tissues, the study of cellular physiology and metabolism, and the growth of normal and transformed human cells *in vitro* (Hayflick and Moorehead, 1961; Leibovitz, 1963; Puck and Marcus, 1955).

Arising out of this work was the demonstration that cells in culture could be established as cell lines that maintained, at least in part, the differentiated functions characteristic of their cell type of origin. Thus, the creation of cell lines that maintained some functional properties of adrenal cells, pituitary cells (Bounassisi *et al.,* 1962), neurons (Augusti-Tocco and Sato, 1969), myocytes (Yaffe, 1968), and hepatocytes (Thompson *et al.,* 1966) allowed the study not only of growth but of the response to hormones and other environmental factors and the production and secretion of hormones and other differentiated functions *in vitro*.

The demonstration that each cell type has an optimal mix of nutrients that supports its function (Ham and McKeehan, 1979; Waymouth, 1981) has led to media derived to support specific cell types under specialized conditions. In parallel, the recognition that serum could be replaced by defined components such as attachment proteins, transport proteins, and hormones and growth factors (Barnes and Sato, 1980a,b; Bottenstein *et al.,* 1979; Mather and Sato, 1979) once again opened up new possibilities for the maintenance of specialized cells and tissues in culture, and thus the ability to address important biological questions in new ways.

Finally, the advent of recombinant expression in mammalian cells and the creation of antibody-producing hybridoma cell lines, coupled with the use of large-scale culture techniques for culturing mammalian cells, has created an important niche for industrial cell culture as a production system for recombinant proteins. The special considerations inherent in industrial production using large-scale cultures have further increased our understanding

Figure 1.2. A photograph of an industrial large-scale culture facility used to make recombinant proteins in mammalian cell culture.

of the range of cell "behaviors," their inherent stability, the ability to genetically manipulate cell properties, and the technical challenges of growing mammalian cells in tanks large enough to have several atmospheres' difference in pressure from top to bottom (Fig. 1.2).

Each of these insights and technical advances has brought new challenges, raised more questions, and widened our experience with that "microorganism" (see Puck, 1972), perhaps better defined as a "social organism," which is the mammalian cell *in vitro*.

THE PRACTICE AND THEORY
OF TISSUE CULTURE

This book is meant to serve as an introduction to cell culture both for students who have little or no experience of cell culture and for scientists who do have some experience with sterile technique and mammalian cell culture and wish to set up a cell culture facility in their laboratory. Thus, each section on the techniques, space, and equipment will be divided into a "minimal," "standard," and "optimal," or "ideal" laboratory. The minimal facility is described as one that can be used for a teaching laboratory or for a laboratory where there is only an occasional use for tissue culture. The standard facility should be considered the desired level if tissue culture is an important and frequently used part of the research work (e.g., a laboratory that studies expression of recombinant proteins) but is not the central task of the laboratory. The optimal facility described is one that should be achieved if cell culture is of critical importance to the work done in the laboratory (e.g., new cell line development, *in vitro* studies of the regulation of gene function, etc.). One can, of course, mix equipment and space considerations based on the available space, equipment, and research goals.

In parallel, the book will cover the concepts and technology of cell and tissue culture on several different levels. Cell culture consists of a few basic concepts and techniques that can and should be mastered at the student or introductory level in a few weeks or months. These include sterile technique, subculture of cells, freezing and thawing cells, cloning cells, measuring cell growth and viability, and starting primary cultures. With these few techniques the scientist or student can usually successfully handle many of the experiments performed with established cell lines, especially those lines that are relatively hardy.

However, it is important for the scientist who makes tissue culture an important tool in his or her research to have a more complete understanding of the science and the years of experimentation behind these techniques. What does the medium do for the cells? How does the choice of incubator setting and medium interrelate? How can the environment be altered for optimal growth of cells at high density? How should the medium be changed for suspension culture? What does one do when the cells "just die"? There is an extensive body of information available that will help answer these questions; however, far too many scientists are content to ask someone else when they have a problem and consider the solution "magic." As flattering as it may be to be considered a magician, it is by far preferable that each person doing cell culture have a good basic understanding of the principles behind the subject. We will attempt to discuss these basic principles in this book.

Finally, at the third level we will give an introduction to some of the techniques that may be used by only a few scientists, but which begin to demonstrate the full power and flexibility of the technology and provide an understanding of *in vitro* cell biology as an approach to answering some of the most basic questions in science. Various approaches to specific scientific problems will be mentioned, with the emphasis on understanding when to select which approach or technique. We will then refer to the literature for detailed and more extensive descriptions of specialized techniques. In most cases, entire methods books are available devoted to a single topic such as expression of recombinant proteins or culture of neuronal, liver, or endothelial cells. In the space available, this book can only attempt to direct the reader to the appropriate references for further reading.

Likewise, while we have attempted to provide an appendix containing lists of vendors and sources for supplies and equipment (including Internet addresses) that are sufficiently complete to allow one to find all of the materials described here, these lists are by no means exhaustive. Researchers may find another source for any one of these materials or alternative equipment that is of good quality and perhaps better suited to their needs, budget, or locale.

PRIMARY CULTURE

Several different types of culture are routinely performed. These can be roughly divided into "primary culture" and "culture of established cell lines." Primary culture can consist of the culture of a complex organ or tissue slice, a defined mixture of cells, or highly purified cells isolated directly from the organism, as illustrated in Fig. 1.3A, B, or C. More commonly, techniques may be employed to purify the cell type of interest and start a primary culture consisting largely of that one cell type. Such cultures usually start at initial plating as containing 60–95% of the cell type of interest, although this percentage may increase or decrease during the subsequent culture period. However, primary cell and organ cultures have an advantage in that they are recently removed from the *in vivo* situation and might therefore be expected to more closely resemble the function of that cell or tissue *in*

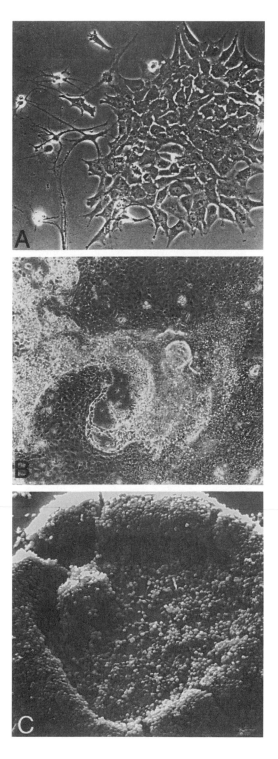

Figure 1.3. (A) Primary culture of isolated granulosa cells; (B) a coculture of several cell types from neonatal rat lung; and (C) a follicle derived from coculture of rat immature granulosa cells and oocytes that reassociate *in vitro*.

vivo. The disadvantage is that these cultures are reacting to a constantly changing environment over the first days or weeks *in vitro,* including the damage sustained during the removal of cells from the animal and tissue and partial recovery from this damage, the change in environment from the animal to the *in vitro* culture, and the changing composition of the culture as some cells in the mixed cultures die and others proliferate and/or differentiate.

ESTABLISHED CELL LINES

The second type of cell culture is the culture of established or immortal cell lines. The vast majority of these are derived from tumors (e.g., HeLa) or from cells transformed *in vitro,* although some of the very earliest lines were established from normal embryonic tissue (e.g., 3T3, CHO). There are also lines that have been widely used, such as WI-38, which are from normal human tissue and have a limited life span *in vitro.*

These cell lines have been the workhorses of cell culture, from their use in studying the control of the cell cycle to vaccine production and large-scale industrial production of recombinant proteins in 12,000 liter tanks. Not surprisingly, after many decades of growth in many laboratories they are both relatively tough (i.e., resistant to temporary lapses in good cell culture technique) and have altered from their original phenotype. Thus, cells having the same designation carried in different laboratories may vary considerably in their properties. We will use some of these commonly available cell lines for the exercises described, but some variation in response is to be expected when cells are obtained from different laboratories.

More recently, cell lines have been developed with the aim of maintaining a normal phenotype combined with the ability to grow the cell, or its precursor, indefinitely in culture. This can be accomplished using conditional transformation or by establishing the cell line from stem cell or precursor cells, which can then be induced to differentiate into a terminally differentiated cell type in culture. These lines are generally more challenging to handle *in vitro* and will be covered in the last section.

THE PHYSICAL AND CHEMICAL ENVIRONMENT

Basically, the aim of mammalian (or any other) cell culture is to provide an environment that mimics, to the greatest extent possible, the *in vivo* environment of that specific cell type. The cell culture incubator, the culture dish or apparatus, and the medium together create this environment *in vitro.* They provide an appropriate temperature, pH, oxygen, and CO_2 supply, surface for cell attachment, nutrient and vitamin supply, protection from toxic agents, and the hormones and growth factors that control the cell's state of growth and differentiation. Clearly, this is not a simple system. In past years, the very process of defining media and culture conditions for cells has increased our understanding of how the cells and the organism from which they come function. The continuing refinement of these conditions to allow the growth of cells not previously cultured *in vitro* (Li *et al.,* 1996; Loo *et al.,* 1989; Roberts *et al.,* 1990) or the maintenance of a complex phenotype *in vitro* (Li *et al.,* 1995) continue to inform us of the cell's needs, interactions, and associations in its *in vivo* state. Thus, cell culture continues to be not just a tool but also a window into the *in vivo* environments of each cell type studied *in vitro.*

The pioneers of cell culture used exceedingly complex environments to maintain their pieces of tissue or cells *in vitro,* including plasma clots and tissue extracts in complex hand-blown glass chambers designed to maintain sterility while providing gas exchange (Fig. 1.1). Then, with the use of better and more complex media, serum supplementation alone was sufficient to grow many transformed cell types in culture, although other cells still required "feeder" (Puck and Marcus, 1955) layers of other cells that were treated so as to be unable to proliferate but provided necessary nutrients or substrates.

Advances in our understanding of what is important in these complex mixtures has led to an increased ability to simplify and define the growth conditions and tailor them specifically to the task at hand. Thus, controlled conditions and minimal cost may be most important in producing large amounts of recombinant protein. In contrast, providing the appropriate defined substrate and growth factors may be critical to maintaining differentiated function or regulating differentiation in experiments designed to study these processes. Understanding the basic theoretical cell culture framework allows one to tailor the culture system to provide the desired outcome. In the last 15 years, the development of serum-free, defined growth media for a number of different cell culture systems and the commercial availability of many purified reagents used in these cultures have aided in these and other endeavors.

The use of more defined media and growth factor supplements has also highlighted the role of the substrate to which the cells are attached in regulating growth and differentiated function. These attachment factors, such as collagen, fibronectin, and laminin, are part of the complex *in vivo* environment in which a cell normally functions. For some cells the cell shape per se is also an important factor in how the cell functions. Complex materials are available to control cell shape, spreading, and attachment and even allow the reproduction of mechanisms that control cell stretch, in order to study this phenomenon *in vitro.* Having defined the components of the medium, the hormones and growth factors, and the attachment factors, we can look again at the complex ways in which two cells interact *in vitro* with a new level of understanding.

FURTHER INFORMATION

Finally, we hope to provide useful hints on sources of materials and equipment and references to both the primary literature and other methods volumes that describe specialized techniques and specific areas of interest in more detail. For example, we will briefly mention how to recognize and measure apoptosis, or programmed cell death, as it occurs in cell cultures and how to differentiate it from necrotic cell death. We will not go into great detail on the very extensive literature on the subject or the many complex methods of measuring apoptotic cell death, but rather we will provide references to the volumes available on this topic. In parallel, we might suggest cheaper alternatives to standard equipment for use in the classroom but will assume that most readers of this book will wish to use commercially available equipment and supplies rather than putting together or building their own equipment or purchasing extremely expensive, albeit faster (or bigger or more sensitive), equipment that does the same job.

INTRODUCTION

Agusti-Tocco, G., and Sato, G., 1969, Establishment of functional clonal lines of neurons from mouse neuroblastoma, *Proc. Natl. Acad. Sci. USA* **64:**311–315.

Barnes, D., and Sato, G., 1980a, Methods for growth of cultured cells in serum-free medium [review] [65 refs], *Anal. Biochem.* **102:**255–270.

Barnes, D., and Sato, G., 1980b, Serum-free cell culture: A unifying approach, *Cell* **22:**649–655.

Bottenstein, J., Hayashi, I., Hutchings, S. H., Masui, H., Mather, J., McClure, D. B., Chasa, S., Rizzino, A., Sato, G., Serrero, G., Wolfe, R., and Wu, R., 1979, The growth of cells in serum free hormone supplemented media, *Methods Enzymol.* **58:**94–109.

Bounassisi, V., Sato, G., and Cohen, A., 1962, Hormone-producing cultures of adrenal and pituitary tumor origin, *Proc. Natl. Acad. Sci. USA* **48:**1184–1190.

Carrel, A., 1912, On the permanent life of tissues outside of the organism, *J. Exp. Med.* **15:**516–528.

Eagle, H., 1955, Nutrition needs of mammalian cells in tissue culture, *Science* **122:**501–504.

Ham, R. G., and McKeehan, W. L., 1979, Media and growth requirements, *Methods Enzymol.* **58:**44–93.

Harrison, R., 1907, Observations on the living developing nerve fiber, *Proc. Soc. Exp. Biol. Med.* **4:**140–143.

Hayflick, L., and Moorehead, P., 1961, The serial cultivation of human diploid cell strains, *Exp. Cell Res.* **25:**585–621.

Leibovitz, A., 1963, The growth and maintenance of tissue-cell cultures in free gas exchange with the atmosphere, *Am. J. Hyg.* **78:**173–180.

Li, R., Phillips, D. M., and Mather, J. P., 1995, Activin promotes ovarian follicle development *in vitro, Endocrinology* **136:**849–856.

Li, R. H., Gao, W.-Q., and Mather, J. P., 1996, Multiple factors control the proliferation and differentiation of rat early embryonic (day 9) neuroepithelial cells, *Endocrine* **5:**205–217.

Loo, D., Rawson, C., Helmrich, A., and Barnes, D., 1989, Serum-free mouse embryo cells: Growth responses *in vitro, J. Cell. Physiol.* **139:**484–491.

Mather, J. P., and Sato, G. H., 1979, The use of hormone-supplemented serum-free media in primary cultures, *Exp. Cell Res.* **124:**215–221.

Puck, T., 1972, *The Mammalian Cell as a Microorganism: Genetic and Biochemical Studies in Vitro,* Holden-Day, San Francisco.

Puck, T., and Marcus, P., 1955, A rapid method for viable cell titration and clone production with HeLa cells in tissue culture: The use of x-irradiated cells to supply conditioning factors, *Proc. Natl. Acad. Sci. USA* **41:**432–437.

Roberts, P. E., Phillips, D. M., and Mather, J. M., 1990, Properties of a novel epithelial cell from immature rat lung: Establishment and maintenance of the differentiated phenotype, *Am. J. Physiol. Lung Cell. Mol. Physiol.* **3:**415–425.

Sanford, K., Earle, W., and Likely, G., 1948, The growth *in vitro* of single isolated tissue cells, *J. Natl. Cancer Inst.* **9:**229–246.

Thompson, E., Tompkins, G., and Curran, J., 1966, Induction of tyrosine α-ketoglutarate transaminase by steroid hormones in a newly established tissue culture cell line, *Proc. Natl. Acad. Sci. USA* **56:**296–303.

Waymouth, C., 1981, Major ions, buffer systems, pH, osmolality, and water quality, in: *The Growth Requirements of Vertebrate Cells in Vitro* (C. Waymouth, R. Ham, and P. Chapple, eds.), pp. 105–117, Cambridge University Press, New York.

Yaffe, D., 1968, Retention of differentiation potentialities during prolonged cultivation of myogenic cells, *Proc. Natl. Acad. Sci. USA* **61:**477–483.

CHAPTER 2

Setting Up a Cell Culture Laboratory

SPACE REQUIREMENTS

Ideally, the space allocated for the tissue culture laboratory should be one dedicated to tissue culture functions exclusively, in order to minimize the introduction of potential contaminants. Traffic in and out of the culture room (space) and talking in the space are to be discouraged. All tasks that do not need to be performed in the culture room (e.g., which do not require a sterile environment) should be performed elsewhere. People not actively engaged in doing cell culture should leave the room. Minimize entry and exit, for example, by having a refrigerator and freezer in the culture room or an airlock "entry room" so that there is no need to enter and exit the culture room during the course of an experiment to obtain reagents necessary for the culture work. If space allows, an airlock can help to ensure a "clean" tissue culture room. If it is not possible to have a separate room for the cell culture equipment, select a corner of the laboratory that is farthest away from doors and other heavily trafficked areas. Place all the culture equipment together in this area of the room and remove any equipment not needed for cell culture to another area of the laboratory. This area should then be cleaned and maintained as described.

The need to minimize the potential for contamination requires that the room be kept under positive pressure with high-efficiency particulate air (HEPA) filtered air flowing through it. The floors should be smooth and untextured. If a vinyl floor covering is used, it should be a continuous unseamed sheet. False ceilings are also a potential source of contamination and should not be used with a positive pressure air flow. If possible, a solid ceiling should be constructed. Minimally, new ceiling tiles should be installed every few years (if you see stained or damaged tiles or mold growing on or between the tiles, it is time!) and the space above well cleaned, with any leaks from the outside or from condensation fixed immediately. Plumbing and all other "bulkhead" fittings and hardware should be well sealed where they pass through the wall or ceiling.

Given that there are different needs, depending on the level of tissue culture being done, we will discuss the requirements for three types of tissue culture setups: the teaching laboratory, the standard research laboratory, and the optimal tissue culture laboratory. In all cases, however, the laboratory or designated area should be designed with minimal and optimal flow of traffic. This concept will affect placement of incubators, hoods, microscopes, freezers and refrigerators, and storage of sterile, disposable supplies, as well as positive and negative interaction of personnel. Figs. 2.1–2.4 show floor plans for several different tissue culture laboratory configurations, depending on space limitations, funding, and usage level.

If you have the opportunity to design and construct or renovate your own tissue culture laboratory space, plan for plenty of "unused" space. This allows for easier cleaning, easier access to equipment, and the ability to add more equipment without costly new construction. There should be one biosafety hood for each person who is a full-time culture room user.

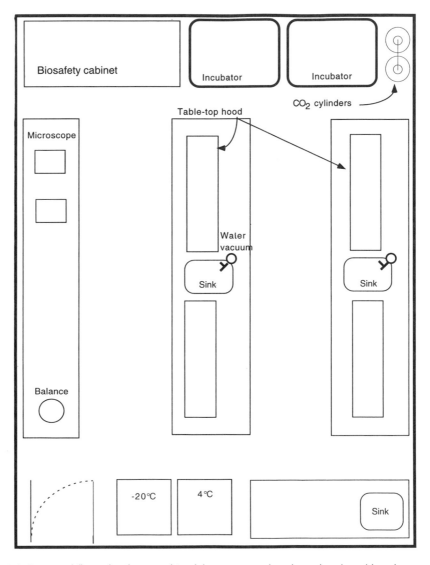

Figure 2.1. Suggested floor plan for a teaching laboratory uses bench top hoods and bench top or free-standing incubators.

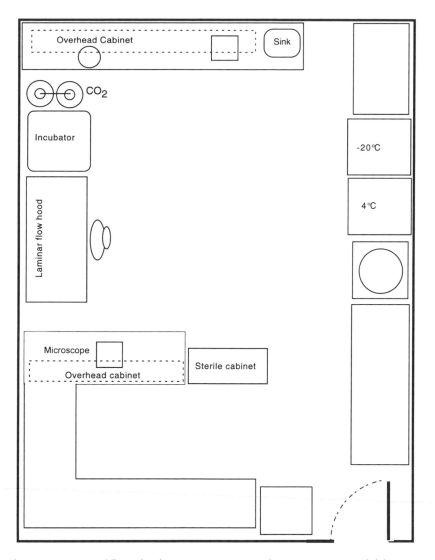

Figure 2.2. Suggested floor plan for setting up a tissue culture area in a research laboratory.

EQUIPMENT

THE TEACHING LABORATORY

The teaching laboratory is a special situation, as it is often a place that is temporary at best and plagued by severe budgetary constraints. The goal here is usually to maintain as aseptic an area as possible for a relatively short period of time. The laboratory and equipment described below are the minimum necessary to teach students all the basic techniques necessary to do cell culture. While the equipment is fairly rudimentary, the concepts taught need not be. The pioneers of cell culture had even less to work with. Depending on the specific aims of the course, the teaching laboratory will minimally require a refrigerator and

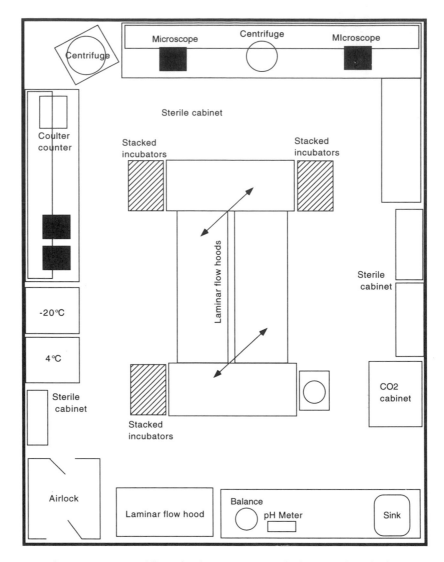

Figure 2.3. Suggested floor plan for a separate standard tissue culture facility.

an incubator, although this need not be a CO_2 incubator, as there are commercially avail-able, appropriately buffered media that obviate the need for CO_2. In this case, the incuba-tor need only be able to accurately maintain the required temperature.

An inverted phase contrast microscope to be shared by all students equipped minimally with a 10x and preferably also a 20x objective should prove adequate for nearly all work done in a teaching laboratory. Most manufacturers offer an economical alternative to their high-end microscopes (Fig. 2.5).

Figure 2.1 illustrates how to set up a standard teaching laboratory for tissue culture us-ing bench top hoods and CO_2 cylinders. This setup requires a minimum of construction. A biosafety cabinet or tissue culture hood, while useful, may not be absolutely necessary in the teaching laboratory, especially if the facility is only temporary or being used for other laboratory functions. We have had success teaching students to work on open bench tops using flasks for cell culture. The medium can be equilibrated for 5% CO_2 by blowing into the flask using a cotton-plugged pipette, or flushing with an air–5% CO_2 mixture from a commercially available gas supplier. Commercially available acrylic bench top hoods are satisfactory, but even this may be unnecessary if the immediate environment is relatively clean, the work area adequate (large enough and of a nonporous material so that it can be cleaned) and disinfected prior to use, and appropriate antibiotics used in a judicious manner. Minimally, the surface on which the culture work is to be done should be nonporous (e.g., stainless steel, Formica, or composite) and cleaned with a disinfectant **before** and **after** each use.

THE STANDARD TISSUE CULTURE LABORATORY

This may be a core facility, working in conjunction with other laboratories, or it may be the primary facility for a cell biology laboratory. All basic cell culture studies can be done in this facility that would be adequate for cell culture use for most cell and molecular biology laboratories, as well as use by physiologists, biochemists, and others who might occasionally need access to a culture facility. If it is to be a shared core facility, the need for min-

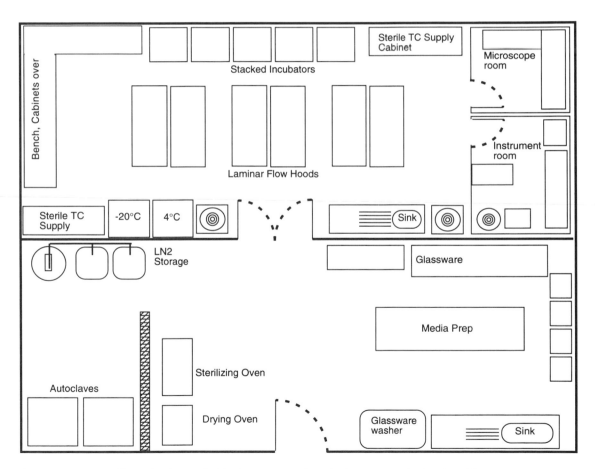

Figure 2.4. Suggested floor plan for an ideal high-use tissue culture facility or a core facility shared by a number of laboratories.

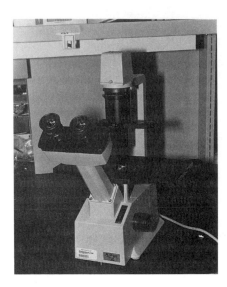

Figure 2.5. An inexpensive inverted phase contrast microscope suitable for a student laboratory, or as an extra microscope in a high-use laboratory.

imizing potential contamination vectors becomes paramount. This is critical not only with regard to bacterial or fungal contamination, but also with cross-contamination of cell cultures. Some facilities have an airlock that can serve as a buffer zone and somewhat of a deterrent to bacteria–fungal contaminants, largely because in its most minimal configuration, it discourages unnecessary traffic in and out of the primary laboratory (see Fig. 2.3). Optimally, the entryway might be large enough to include a sink for hand washing, storage for sterile tissue culture supplies, and even space for a freezer or cell counter. A high-use tissue culture core facility might be designed around a plan such as that shown in Fig. 2.4.

When this is not possible, or is impractical, the laboratory should be designed so that cell cultures can be "compartmentalized," that is, primary cell cultures can be handled in a specifically designated hood(s), and kept in an incubator chamber separate from other cell lines being maintained in long-term culture. Cell cultures coming into the laboratory, as frozen vials or as viable cultures, primaries, or established cell lines, should be quarantined in an incubator chamber and handled in a designated hood until they are tested for mycoplasma.

Incubators are available as two-gas (CO_2–air) or three-gas models (CO_2–O_2–NO_2), this being largely determined by cost and specific needs of the investigators. It is always possible to augment a two-gas incubator chamber when this has not been incorporated into the original design. Insulation is maintained by either a water jacket or an air jacket, with corresponding advantages and disadvantages. The water-jacketed incubator can maintain temperature over a longer period of time should there be a power outage, and this can be a critical feature for some installations. It is much heavier when filled, however, and the level must be maintained by periodic "topping off" and the jacket drained when the incubator has to be moved. The air-jacketed incubator is lighter, has more moving parts to fail, comes up to temperature faster, but will lose heat much faster when the fan goes off (e.g., in case of an electricity shutoff).

More importantly, the interior design and construction, materials used, and ease of assembly and disassembly can determine in part how well the cultures can be maintained free of contamination. Contaminating mold will grow on stainless steel, labeling tape, and even plastic, so shelving and the hardware securing it must be easy to remove and clean when necessary. Copper shelving and interior walls can inhibit the growth of such organisms but it is expensive, and unless all hardware components are of copper construction, one cannot completely inhibit the growth of mold on interior surfaces. Routine cleaning of stainless steel or aluminum shelves with a disinfectant and ethanol rinse will help to reduce these risks. Be careful to use a disinfectant recommended by the incubator supplier. Many excellent disinfectants are volatile and will kill cultured cells as well as contaminants. The chamber should be allowed to equilibrate overnight after a thorough cleaning prior to returning the cultures.

Incubator manufacturers have various proprietary methods of delivering and regulating gas flow into the chamber. Independent of this, the tissue culture laboratory needs a supply of CO_2 and any other gas that will be delivered to the chamber. In-house supplied gas can be a relatively inexpensive source; it need only be equipped with a miniature regulator to reduce the house gas to a flow rate optimal for the incubator and an in-line 0.2-μm filter to prevent introduction of mold and other potential contaminants that can accumulate on the inner wall of a gas hose. If O_2 experiments are to be carried out in an incubator that has such a provision, then house nitrogen should also be plumbed into the laboratory. Alternatively, when house gas is unavailable, commercial gas cylinders may be used, the critical point being to maintain an uninterrupted flow of gas. A manifold, connecting two or more supply cylinders with two-stage regulators, can be configured to supply the gas efficiently and economically. Reinforced silicon tubing that can be sterilized by autoclaving should be used to connect the gas source to the incubator.

It is important to know if the displays on the incubator control panel are reflecting in fact the actual conditions inside the chamber. For accurate temperature determination, a portable RTD thermometer is recommended. The appropriate thermometer–probe combination can provide accuracy to within 0.05°C. CO_2 can be monitored fairly accurately and inexpensively with a Fyrite unit.

The tissue culture hood can be as simple as an open, laminar flow unit, with air passing initially through a HEPA filter and moving parallel to the work surface, exiting at the front of the hood. However, current regulations may require the use of biosafety cabinets, in which HEPA filtered air is circulated within the hood and exhausted through appropriate filters and ductwork to the outside. These hoods are generally available in 4-ft and 6-ft lengths, the latter being somewhat more convenient in terms of work space. Two people can also work side by side in a 6-ft hood. This is convenient if the experimental protocol requires two people to work together. Larger hoods also may be necessary if large-scale tissue culture work is to be done where many large spinners or roller bottles are handled at the same time.

Regardless of the size being used, it is important that the interior of the hood be as free of obstruction as is practical to optimize airflow. Quite frequently one finds the hood being used as a repository for a variety of tissue culture supplies, vacuum units, personal belongings, and other sundry items, leaving little room for work space, thus minimizing airflow and increasing the everpresent risk of contamination (see Fig. 2.6). Keep only the minimum necessary equipment in each hood, have one set of dedicated equipment and supplies for each hood, and restrict the use of that equipment to the hood (e.g., a tube rack, automatic pipettors, or bulb).

No mouth pipetting should ever be done for tissue culture work. Currently available biosafety cabinets have duplex electrical outlets, convenient for plugging in pipetting aids and gas and vacuum valves, and are equipped with UV fixtures (Fig. 2.7). Figure 2.7 (bottom) shows an inexpensive attachment for the hood that allows the storage of pipettes and pipettors conveniently near, but not actually in the hood. There is little need for gas to sup-

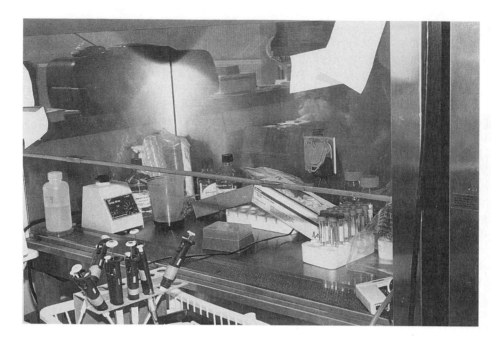

Figure 2.6. A dangerously overcrowded biosafety cabinet. The equipment stored in the hood and on the air intake grid will effectively break the sterile barrier for this hood.

ply a Bunsen burner in this type of hood. In fact, an open flame interrupts the airflow pattern in the hood and decreases barrier efficiency. In addition, an open flame creates a hazardous condition when working with flammable reagents in the hood. Thus, a flame should only be used when essential (e.g., for flaming a coverslip).

The only other item of major importance is a vacuum trap, consisting of two 1- or 2-liter Erlenmeyer flasks connected on one end to the vacuum source (house vacuum or pump) and on the other to a small hook attached to the hood (Fig. 2.8). This apparatus should sit on the floor beneath the hood and be emptied regularly. Tubing, preferably silicon or latex, should have an inner diameter equivalent to the outer diameter of a Pasteur pipette and should be of sufficient length to facilitate aspirating medium and other reagents. The primary flask (at least) should contain disinfectant and the tubing should be flushed with disinfectant after the work in the hood is done. All hoods should have biohazard waste containers lined with autoclave bags.

The laboratory should have at least one inverted phase contrast microscope, equipped with 10x, 20x, and 40x objectives (Fig. 2.9). If fluorescent microscopy will be needed, an epifluorescent attachment should be included. A 4x objective is useful for scanning large fields. For detecting mycoplasma, a 100x objective and fluorescence capability is necessary. A spring-loaded marker that screws into the nosepiece of the microscope is useful for marking areas of interest in culture dishes.

The standard tissue culture laboratory should have a reliable source of water for preparing medium. Considerable study has gone into water quality requirements for optimal cell growth, some cell types being far more sensitive to water quality than others. Nonetheless, all cells respond to water quality and it is important to be able to control this as much as possible. Ideally, water supplied by the city or county should first pass through a deionization unit. Often, institutions will have a source of deionized water supplied to the laborato-

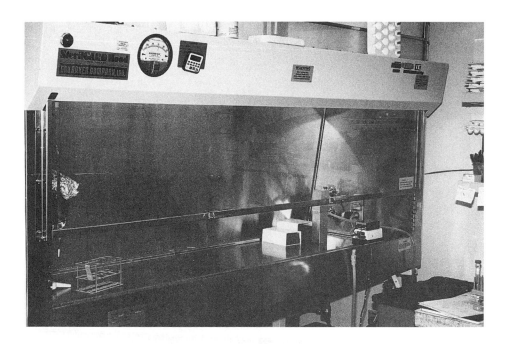

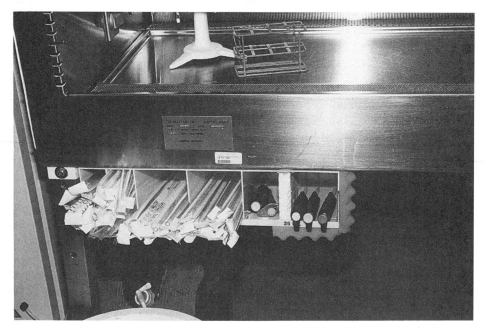

Figure 2.7. A functional biosafety hood. Proper use requires that the hood be kept clean, with a minimum amount of equipment or supplies stored in the hood. This type of cabinet protects the user as well as the cultures by filtering the air coming out of the hood. Lower photo: closeup of hood showing racks for pipettes and pipettors.

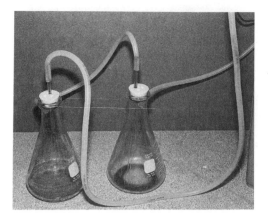

Figure 2.8. Illustration of a vacuum trap and suction setup for medium disposal. The tubing connects the bottle on the left to a vacuum device. The tubing from the bottle on the right is used to evacuate media from plates and tubes by inserting a sterile Pasteur pipette in the end. The end can be hung from the front of the hood with a paper clip when not in use. The tubing from such a device can be seen in the photo of a hood in Fig. 2.7 (upper right).

ries. To this source, the investigator should connect a purification unit, usually in the form of several organic resin cartridges, a charcoal cartridge to remove organic compounds (including those leaching from the previous column), and a final ultrafiltration cartridge (Fig. 2.10), with a 0.2-μm filter attached to the outlet. A still can also be used but this should be a double distillation with potassium permanganate in the first reservoir to remove organic material.

For teaching laboratories using tougher cells (and especially with serum-containing

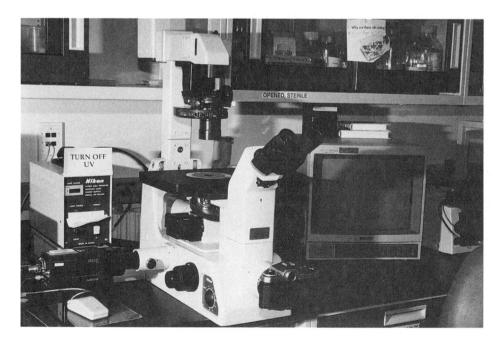

Figure 2.9. Example of a more expensive, high-quality, phase contrast microscope well equipped for viewing cultures using phase contrast or UV fluorescence. There is a video camera attached to the microscope (left port) that allows a frame-grabber to send images for storage on the computer at left (just seen on the left edge of the photo). The camera back, which is attached to the front of the microscope, allows the use of conventional film. The video camera image can also be displayed on the monitor to the right to allow several people to simultaneously view the image.

Figure 2.10. An example of a water purification system that produced water suitable for medium preparation. (Photo courtesy of Millipore Corp.)

medium), bottled spring or distilled water can even be used. The water source should be tested, as some bottled waters are better than others for making media. This is not recommended, however, if a water purification system is available, as there are some cell types that will not grow if water quality is poor. Do not use tap water or water straight from a deionizing column.

Liquid nitrogen freezers for long-term storage of cell cultures should be considered necessary for any standard tissue culture facility. These can be "portable" units, which can be moved when needed and hold up to 2,500 vials. This type of unit is filled with NO_2 manually. Alternatively, there are stationary freezers that can hold up to 10,000 vials or more and are filled automatically by a supplied NO_2 source tank. In either case, but particularly the former, it is absolutely necessary to regularly monitor the liquid nitrogen levels. Cells may be stored frozen at $-80°C$ for a few months but lose viability rapidly at this temperature. Any facility that intends long-term storage of cultured cells should have liquid nitrogen storage capability, preferably a tank that stores cells in the vapor phase, since this minimizes the possibility of cross-contamination of cells during storage (see Chapter 7).

For the student laboratory, cells can be obtained from a friendly research laboratory. Plate the cells at a relatively low density in 25-cm^2 flasks (10^4–10^5 cells/flask). Give a flask to each group of students. Prepare backup flasks, flush the flask with 5% CO_2, and leave them sitting in a dark place at room temperature. Many cells will remain viable under these conditions for several weeks, although growth is extremely slow. If these need to be used, they should be placed in the incubator and grown to confluence before subculture.

THE OPTIMAL TISSUE CULTURE LABORATORY

The optimal tissue culture laboratory is one that would serve a scientist whose main focus of research is cell biology. Thus, there would be multiple laboratory members using the facility and this use would go beyond merely growing cells to be used in experiments (see Chapter 10 on establishing new cell lines). The researchers might be studying many different aspects of cell behavior and working out improvements in the technical aspects of cell culture. This setup would necessarily include all of the above-mentioned equipment as well as additional equipment for visualization and maintenance of cell cultures (for example, Fig. 2.4). This could include multiple gas incubators, microscopes with upright/bright field capabilities, as well as a "dedicated" fluorescence microscope, upright or inverted, with objectives optimized for fluorescence. Such a laboratory also might need increased capability for observing cell cultures such as a facility for larger-scale cultures incorporating the use of roller bottles and spinner flasks. In addition, special microscopes for real-time three-dimensional observation and time-lapse videography can provide the investigator with powerful imaging tools. The ability to capture an image directly from the microscope is becoming increasingly important, and this may be done in a number of ways, such as videography of viable cultures over time or the utilization of image capture boards to digitize, display, and analyze a single image. (See the section in Chapter 6 on visualizing cells to determine what type of equipment is needed for your type of work.)

Another consideration in the optimal setup is vibration isolation for microscopes. This becomes critical in situations where tables or benches used for the microscopes are situated adjacent to equipment with heavy and constant power drains: refrigerators, freezers, centrifuges, and so forth. Vibration isolation can be critically important in achieving success with a particular technique—microinjection, for example.

An electronic particle counter, such as the type made by Coulter Electronics, is extremely useful when conducting experiments that require numerous and regular cell counts, although alternative if somewhat indirect methods are becoming more widely available and may be more practical for experiments utilizing multiwell plate formats (see Chapters 5 and 11).

PLASTICWARE AND GLASSWARE

Plasticware, although expensive, is now preferred over glass for pipettes, tissue culture dishes, and so forth. If glassware, such as spinner flasks, is used for suspension cultures, it should be sialated to minimize the chance of cells sticking to the surface. Use of sialated tubes for media collection and storage before assay for secreted factors is also recommended. Do not sialate if cells are to be grown directly on glass surface or on matrix-coated surfaces, for example, coverslips. Glassware used for tissue culture should be autoclaved with moist heat for 30 min at a pressure of 15 lb/in². Empty bottles should have their screw caps loosely on and covered with foil. Partly filled bottles should have their caps tightened. The glass bottles should dry slowly after sterilization to avoid condensation, which would otherwise be trapped inside. This condensation will dilute solutions placed in the bottle and may contaminate them with chemicals (e.g., boron) leached from the glass. Glass pipettes, tubes, and other small glass items can be sterilized in an oven with dry heat for 90 min at 160°C.

Tissue culture plasticware is available in a variety of formats, ranging from Terisaki plates to larger multiwell plates in 96-, 48-, 24-, 12-, and 6-well configurations (see Fig. 2.11 and Table 2.1). In addition, chamber slides in different configurations are available, in which the upper chamber is removed so the slide can be used in immunocytochemistry or for viewing on an upright bright field microscope. Tissue culture flasks are available with "vented" caps that have filters incorporated in them for gas exchange without having to loosen the cap itself. Culture vessels useful for directly scaling up attached cultures are available. These include roller bottles, large stacked surfaces, hollow fibers, and so forth. They are listed in Appendix 5 and discussed in Chapter 11 on special growth considerations.

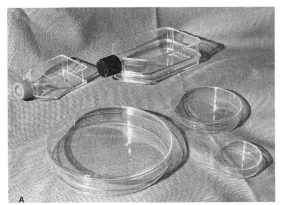

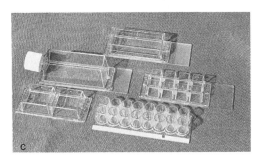

Figure 2.11. A selection of the many types of sterile disposable tissue culture dishes available. (A) The most commonly used flask and dish sizes. (B) Multiwell plates are easily handled and commonly used for higher throughput assays. The Terasaki plate (upper row, center) is used for very-small-volume cultures. (C) Culture slides such as these can be used to grow cells that will be used for microscopy, staining, immunohistochemistry, *in situ* mRNA localization, and so forth.

MAINTAINING THE LABORATORY

Routine maintenance is essential to having a functional cell culture laboratory. While one or more persons may be designated to oversee this task, all users must participate to some extent. This section will deal with those tasks necessary to keep a cell culture lab in good working order. Above all, the entire laboratory should be kept neat, clean, and uncluttered. Clutter provides a hiding place for contaminants and makes the space more difficult to use. Provide a designated space for all equipment and supplies and see that these items are returned to that space by all users. Optimally, when the tissue culture room is shared by several frequent users, each will have a designated space for their supplies and experiments in the refrigerator, a drawer, and incubator. Less frequent users might share a common space. It is generally more efficient for the hoods to be used on a "first-come-first-served" basis, although a sign-up sheet may be necessary for large blocks of time or if there are a larger number of users.

DAILY TASKS

The hoods should be wiped with a mild disinfectant such as 70% ethanol before and after each use by each person using the hood. The trash receptacles and the media waste re-

Table 2.1
Tissue Culture Dish Comparison: Diameter, Growth Surface Area, and Maximum
and Recommended Medium Volume

Dish	Diameter (mm)	Growth area (cm^2)	Maximum volume/well (ml)	Recommended volume for growth (ml)
Plate diameter (mm)				
35	35	9.6		2–3
60	60	28.3		3–5
100	100	78.5		8–10
150	150	148		25
245	245×245	500		100–250
Flask				
25		25		5
75		75		10–20
75, ribbed[a]		125		10–50
150		150		25–50
150, ribbed[a]		265		25–100
175		175		50–100
225		225		50–100
Multiwell plate		Area/well (plate), cm^2		
6	34.6	9.4 (56)	16.8	2.0–4.0
12	22.1	3.83 (46)	6.9	1.0–3.0
24	15.5	1.88 (45)	3.4	0.5–2.0
48	11.6	1.0	1.6	0.25–1.0
96	6.4	0.32 ((31)	0.37	0.05–0.1
Roller bottles				
1 liter		490	1300	50–100
2 liter		850	2200	100–200
Expanded[a]		1700	2200	100–200

[a]Corning offers an "expanded surface" that consists of a ribbed or pleated surface.

ceptacles should be taken outside the tissue culture room and emptied at the end of the day, or more often if they are full. If any contaminated cultures have been discarded, they should be removed from the room immediately and autoclaved before disposal. In most areas, biosafety rules require that tissue culture waste be treated as a biohazard. Dry waste that has come in contact with cells (e.g., empty tissue culture dishes) can be discarded into biohazard bags and autoclaved. Liquid waste from the traps should be disinfected by the bleach or disinfectant in the trap. If there is excess medium, add more disinfectant before disposal. If any medium has been spilled in the hood, the trash, or on the floor, this should be cleaned up and the area disinfected immediately. Media are designed to grow cells, including prokaryotic cells, and do so just as well when spilled.

The hood sashes should be placed at the indicated level and the UV lights turned on in the hoods at the end of each use. Do not turn hoods off overnight, as continuous operation of the blowers and germicidal lamps will aid in maintaining aseptic conditions in the hood. Microscopes, cell counters, and other small equipment should be turned off. Visually check the temperature and CO_2 readouts on the incubators to confirm that they are operating within acceptable tolerances. All of the above should be rechecked at the end of the day by the last person leaving the room. Hoods, benches, microscopes, and water baths should be checked to make sure no reagents or cultures have been left out.

The liquid nitrogen level should be checked in all storage tanks. The incubators should be checked for water. Sterile supply cabinets should be restocked. For serum-free culture work, it is imperative to have growth factors, hormones, and other required components on hand at all times. It may be necessary to maintain a list of these components on the outside of the refrigerator or freezer where the items to be ordered can be marked accordingly.

MONTHLY TASKS

These tasks may be performed less often if the facility is not heavily used, but all should be done at least annually. The room should be thoroughly cleaned, including the floors and bench tops. Ask the cleaning staff to use a new mop and fresh cleaning solution to clean the floor. The incubators should be disassembled, the removable parts autoclaved, and the rest wiped with disinfectant. The CO_2 levels and temperature levels should be checked and calibrated. If a calibration service is available, have the incubators calibrated at least once a year and preferably every 6 months. In lieu of this service, check the temperature with an RTD thermometer and the CO_2 with a Fyrite unit on a regular basis. The hoods should also be cleared out, disassembled (just the work area), and all surfaces wiped with a strong disinfectant, such as Betadine (povidone-iodine). All automatic pipettors should be checked for accuracy. Weighing the same nominal volume from several pipettors is a quick method to check accuracy. Coulter counters and other major equipment should be serviced. Check the water purification unit and clean stills or replace cartridges on the filtration units. Clean out refrigerators and freezers. Throw out any expired solutions and check carefully that there is no mold growing on tubes, tape, or caps of containers in the refrigerator. Microscopes should be cleaned and the alignment checked. UV bulb life should be checked. Check to see that microscope bulbs and other supplies are well stocked.

Contamination and faulty equipment can destroy months of work overnight. The time and attention required for a well-maintained laboratory will repay the investigator with years of trouble-free service.

CHAPTER 3

The Physical Environment

The essential idea of the entire panoply of the chemical solution and mechanical apparatus surrounding the cell in *in vitro* culture is to recreate the physical, nutritional, and hormonal environment of the cell *in vivo*. This includes controlling the temperature, pH, osmolality, and gaseous environment; providing a supporting surface; and protecting the cell from chemical, physical, and mechanical stress. The mammalian body has evolved to do this over billions of years. We are still learning the correct requirements for different cells to function optimally and normally *in vitro*. This chapter will deal with the control of the physical environment. Chapter 4 on media and the section on serum-free media will deal with the nutrient and hormonal environment. It should be recognized, however, that these are interrelated and by necessity inseparable. Thus, the level of nutrients present determines the response to hormones, as does the surface on which the cell is grown. The energy source supplied may alter the osmolality and the temperature will alter the rate of energy use. Again, an understanding of how these basic elements interrelate will allow the cell biologist to alter and even design cell culture systems that are optimal for a specific function, whether supporting the function of a highly differentiated cell such as a neuron or providing the simplest environment possible for the production of recombinant proteins. We will deal with each of the aspects of the physical environment separately.

TEMPERATURE

Most mammalian cell cultures are grown in incubators that are set at 37°C. This was chosen because it is the core body temperature of *Homo sapiens*. One might therefore expect it to be the optimal temperature for growing many cell lines derived from human tissues. However, there is no reason to expect 37°C to be optimal for all types of human cells, and certainly not for all cells from all mammals, although most cells from warm-blooded animals will grow at this temperature. For example, the normal temperature of the human skin is well below the 37°C of the core of the body. Thus, it might be considered more "normal" to grow human epidermal cells at a lower temperature. Likewise, the testis of all mammalian species are maintained at a temperature several degrees lower than the core body temperature (33°C for humans), and spermatogenesis will not proceed normally at core body temperature. It is therefore necessary to use a lower temperature if one wishes to create an *in vitro* environment appropriate for studying all aspects of the process of spermatogenesis.

In contrast, many animals used to derive cell lines (e.g., cows, cats, rats, mice, and hamsters) have a body temperature significantly different than that of humans. There is no *a priori* reason that cell lines from these animals' tissues should be grown at 37°C. However, if the line has been in common use for many years, the cells might be best adapted to this temperature. Insect, worm, and fish cells, on the other hand, cannot be grown at 37°C and require significantly lower temperatures.

There are several general statements about choosing the temperature set point for the incubator. Figure 3.1 shows packed cell volume (a measure of total cell number) and protein secretion by CHO cells grown at various temperatures. Growing cells at too high a temperature is more detrimental than too low a temperature. Even fairly short exposures to temperatures as little as 3°C above normal may lead to the death of the entire culture. Cells can tolerate temperature 10–20 degrees below normal for fairly long periods of time, but will grow slowly, if at all, at these low temperatures. If it is done correctly, cells can of course be frozen to extremely low temperatures of −80°C to −130°C and survive for years in cryostorage. Freezing, thawing, and storage of cells are discussed in Chapter 5.

If studies are to be done comparing properties of cells grown at different temperatures, the two stock cultures should be grown at the two different temperatures for at least one subculture before setting up the comparison experiment. Abrupt changes of temperature

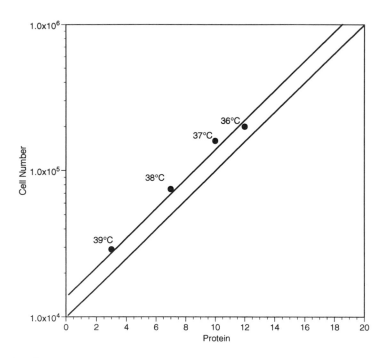

Figure 3.1. The effect of temperature on CHO cell growth and recombinant protein yield *in vitro.* The cells were counted and secreted protein measured after 4 days in culture at the indicated temperatures.

cause many changes in the cells that may take some time to disappear. Likewise, when observing cells in the microscope, do not leave the plates out for long periods allowing them to cool to room temperature as this will affect the growth rate and other properties of the cells. Removing a few plates at a time for observation and returning them immediately after minimizes cooling and the resulting condensation that also changes the properties of the medium.

The primary mode of maintaining temperature in cell culture is the incubator. While most cell culture incubators are set up to regulate humidity and the gaseous environment, as well as control temperature, temperature control is probably the most important part of this since there are alternative methods for controlling pH and humidity. As discussed in Chapter 2, there are several commercially available types of incubators. Generally, incubators that maintain temperature by warming and circulating heated air are more responsive to temperature changes and will restore the temperature faster after the door of the incubator is opened and closed. However, in the event of a power failure or mechanical breakdown, the water-jacketed incubators will maintain the temperature for several hours after power failure. The water also acts as insulation and the circulation of air only within the incubator makes maintaining a humidified atmosphere easier.

In student laboratories, if cell culture incubators are not available, to maintain humidity cells can be grown in stoppered flasks that have been equilibrated with air exhaled from the lungs through a cotton plugged pipette to maintain the 5% CO_2–air mixture required for many media. These flasks can then be placed in a well-insulated box with a heat source and a thermostat. Many cell lines will grow quite well in these conditions in the student laboratory. The thermostat should be able to maintain the temperature set point to within 20°C; the tighter the range, the better. It is possible to have students build an adequate incubator for a minimal amount of money using acrylic, some form of insulation, a light bulb, and a thermostat. Periodically, independently check the incubator set point with a thermometer to make sure the readout is correct.

Maintaining temperature for experiments where the cells cannot be placed in the incubator, such as time-lapse photography, requires special arrangements (see, for example, Fig. 6.17). The microscope stage or other surface the dishes are resting on can be heated or the air around the culture container can be heated by enclosing the microscope stage in an incubator box (see Chapter 6). In large-scale culture, the medium can be directly heated with heating coils or in a reservoir and circulated through the culture device. The cultures can also be placed in a "warm room." These special arrangements will be discussed in the appropriate sections. In all cases, the better the temperature control, the better and more reproducibly the cells will behave.

It is also important to be sure that the heating does not lead to undue evaporation of the medium. In an incubator this is done by maintaining a saturating humidity in the air in the incubator. Air blown over cultures to keep them warm should also have saturated humidity, and cultures kept on a heated surface will work best if the cells are in a sealed flask to minimize evaporation.

Since incubators, hoods, microscopes, freezers, and other equipment in the tissue culture room generate a good deal of heat, there should be good air turnover and a good cooling system for a culture room. If the temperature in the room exceeds 37°C by very much for any length of time, it is possible that all the cultures will be lost, since most tissue culture incubators do not cool. If nonmammalian cell lines, such as those from insects, fish, or amphibians, are being grown in a laboratory, an incubator that will cool the cultures is desirable, since these cell types must be grown at temperatures well below 37°C.

Regulation of extracellular and intracellular pH is also essential for survival of individual mammalian cells and of animals as organisms. The pH is not only important for maintaining the appropriate ion balance, but also for maintaining the optimal function of cellular enzymes and for optimal binding of hormones and growth factors to cell surface receptors. Even transient changes in pH can alter cell metabolism and induce the production of heat-shock proteins, a process which can lead to apoptotic cell death. As with the organism, different compartments of the cells can have widely differing pHs, but the cell will maintain the appropriate pH of its subcellular compartments if the external environment is correct. This means not only providing the correct external pH but also the proper membrane components, ions, and ratio of ions, which allow the cells to maintain its internal pH through the integrity of its ion pumps and cell membranes.

Most media strive to achieve and maintain a pH between 7.0 and 7.4, with a median of 7.2. Different cell types may have an optimal pH slightly outside this range, and cells certainly differ widely in their ability to tolerate significant deviations from this level. As with temperature, slow changes in pH are tolerated better than rapid changes that cause apoptotic cell death. Most cells will tolerate a medium in the range of 6.5 to 7.8, but media much outside this range can be lethal. Even within this range, the growth rate and cellular functions vary widely as a function of pH. Cells will release cell-bound growth factors at a pH of 4–5 and remain intact but not viable. pH 2.0 is used to fix cells for some histochemical procedures. Cell membranes are solubilized at very basic pH, leaving basement membrane and cytoskeletal components on the culture vessel surface.

The regulation of pH is done through a variety of buffering systems. Most media use a bicarbonate–CO_2 system as a major component of the buffer system. (The interaction of medium bicarbonate levels and incubator CO_2 systems is discussed in Chapter 4.) Other buffers included in most media formulations are the phosphate buffers. Media may be further supplemented with complex organic buffers or serum, which when present in levels of 5–20% (vol/vol) of the medium provide significant buffering capacity.

An understanding of the metabolism of the cells is essential to be able to derive the best method to control cell pH using the media and buffering systems available. All cells produce lactic acid and CO_2 as a by-product of their energy metabolism. This is necessary for life. Generally, the faster-growing cells will produce more lactic acid than slow-growing or nongrowing cells, but there is great variability among different types of cells. In addition, there is an interdependence between medium composition and both the absolute and relative amounts of lactic acid produced by a given cell. Cells grown in low concentrations of glucose will convert a higher percentage of the glucose into macromolecules used in cell replication and less into lactic acid than the same cells grown in high concentrations of glucose. Lactic acid itself can be taken up and metabolized by some cells. Additionally, the CO_2 produced by the cells will affect the pH through the bicarbonate–CO_2 buffering system, but will also be released into the atmosphere in the culture container at a rate determined by the culture vessel. Thus, a sealed flask will build up high levels of CO_2, while a fermenter with air sparging will strip much of the CO_2 produced by the cells out of the vessel. The actual pH in a culture at a given time is thus determined by the culture configuration, the medium the cells are grown in, the cell type, and the buffering capacity of the medium.

Cells grown in dishes in incubators are passively controlled by the type of dish, medium composition and depth, and the buffers, while cells grown in fermenters or instrumented spinners have active control of pH through the programmed addition of acid or base to maintain the pH at a given set point. This control comes at a cost, however. These added acids and bases result in an increase in salts in the medium and a resulting increase in os-molality, which eventually will become, of itself, limiting for cell growth.

The main buffering system in most media is the CO_2–bicarbonate system. The inter-action of CO_2 derived from the cells or the atmosphere with water leads to a drop in pH de-scribed by the equation below:

$$H_2O + CO_2 = H_2CO_3 = H^+ + HCO_3^-$$

Increasing the bicarbonate concentration neutralizes the effect of increased CO_2 due to the following: $NaHCO_3 = Na^+ + HCO_3^-$. The increased H drives the equation above to the left until equilibrium is reached at pH 7.4 (if the correct bicarbonate–CO_2 ratios are used). The effect on cell growth of altering the pH by mismatched CO_2–bicarbonate concentrations is shown in Fig. 3.2.

High concentrations of bicarbonate in media formulations require higher CO_2 per-centages in the air supply to provide an appropriate pH in the medium but also to provide greater buffering capacity. The carbonate ions themselves also affect cell function to some extent, and some cell types require bicarbonate to grow, independent of its requirement as a buffer. The choice of buffering system will clearly affect both the pH and the osmolality of the final medium, as well as other aspects of cell physiology controlled by the ionic en-vironment.

The bicarbonate buffering system, while inexpensive and well understood, may be supplemented with a variety of organic buffers. The most widely used zwitterionic buffer is HEPES (N-2-hydroxyethylpiperazine-N'-2-ethane sulfonic acid). Since these organic buffers are insensitive to the CO_2 level, they provide a good backup system for rapidly me-tabolizing cells that produce a lot of CO_2, and can stabilize pH swings that occur in CO_2–bi-carbonate-buffered media when removing cells from the incubator for observation under

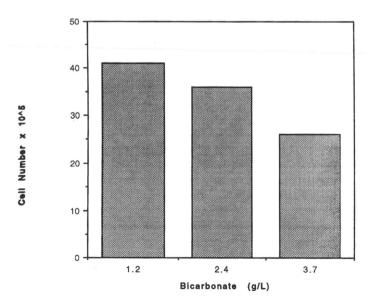

Figure 3.2. Match the CO_2 with the correct bicarbonate level for optimal growth. The figure shows the growth of cells at different bicarbonate levels in a 5% CO_2 incubator. The combination of 1.2 g/liter bi-carbonate and 5% CO_2 provides a pH of 7.2. Increased bicarbonate will raise the pH.

the microscope. This is particularly important when eliminating serum that has a significant buffering capacity itself. Most cells tolerate HEPES buffer added in the range of 10 to 50 mM, or even higher. However, since this is a rather expensive medium additive, it is generally used at 10–20 mM (see Medium Preparation in Chapter 4 for preparing HEPES-buffered medium). Note that many buffers, such as Tris [tris(hydroxymethyl)amino-methane], commonly used in cell-free biochemistry are toxic to cells.

OSMOLALITY

The osmolality of the medium used is determined by the medium formulation. Salts and glucose are the major contributors to the osmolality of the medium, although amino acids may also contribute significantly. Altering the osmolality significantly (by more than 50 mOsm) will almost always affect cell growth and function in some manner. This should be taken into consideration when studying the effects of the addition of ions, energy sources, or large changes in amino acid levels on cell growth and function. Almost all commercial media are formulated to have a final osmolality of around 300 mOsm. While different cells might have somewhat different iso-osmotic points, and therefore a different optimal osmolality for growth (Waymouth, 1970) or a specific function, most cells grow well in the range of 290 to 310 mOsm.

One way to confirm whether media have been properly prepared is to check the osmolality. Significant deviations outside the range of optimal osmolality will result in loss of membrane integrity, as the outside osmotic pressure becomes too much higher or lower than that which must be maintained inside the cell. In short, the cells explode or collapse. This phenomenon can be observed in minutes by removing medium from a dish of cells, replacing it with water, and observing under a microscope. If you wish to study the effects of osmolality on your cells, make up medium without sodium chloride. Add various concentrations of NaCl around the concentration that is normal for the medium you are using and determine the effect on the parameter to be studied (e.g., growth). The effect of high osmolality on cell growth and cell volume is shown in Fig. 3.3.

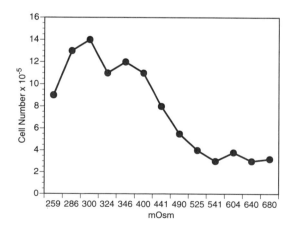

Figure 3.3. The effect of increasing osmolality on cell growth. The optimal osmolality for the growth of these CHO cells is approximately 290 mOsm.

Osmolality can be checked directly with an osmometer. One can also increase the osmolality of medium by adding sodium chloride. Adding 0.0292 g/liter of NaCl will raise the osmolality 1 mOsm (using a 5 M NaCl solution 1 ml/liter will raise the osmolarity 10 mM). This is a rough estimate but useful if a quick experiment is to be performed to measure the effect of increasing osmolality. Remember, in media other components such as sugars, other ions, and amino acids also contribute to the final osmolarity of the medium. Also, be aware that changing osmolality by adding NaCl alters the Na:K ratio, and this may affect other aspects of cell function, including membrane transport. Osmolality may also be increased by adding "inert," that is, nonmetabolizable, sugars such as xylose. While available medium formulations should be optimal for growing most of the commonly used established cell lines, it is advisable to keep the possibility of altering osmolality in mind when trying to culture cell types that are not commonly cultured.

CO$_2$, OXYGEN, AND OTHER GASES

The CO$_2$ level is generally part of the buffering system of the medium and must be set accordingly (see above and "Matching the Incubator Settings and the Medium," Chapter 4). Oxygen levels in the incubator are provided by the 90–95% air in the incubator mix. This means that the PO$_2$ in most incubators is at 18%. To reduce the gaseous oxygen to less than this amount, special incubators must be used that mix CO$_2$, air, and nitrogen to obtain the desired PO$_2$. A cheaper but less precise method is to obtain a sealed plastic box with entry and exit ports and buy premixed gases with the appropriate PO$_2$. The cultures are placed in the box, the lid sealed, the box flushed with the compressed gas mixture through the ports, and the sealed box placed in the incubator to maintain temperature. Be sure to provide a source of moisture, such as an open plate of sterile water, since compressed air has an extremely low moisture content and can cause the medium in the dishes to evaporate, thus raising the osmolality. This method allows several different boxes with different gas mixes to be placed in the same incubator; however, a different tank (or a mixing manifold) will be needed for each mixture. Fermenters can be equipped directly with PO$_2$ sensors that can feed air or oxygen into a tank to maintain a set PO$_2$ level in the medium. This is probably the most accurate way to study the effects of minor variations in PO$_2$ on cell function.

However, most researchers do not deal with direct changes in the PO$_2$ level in the incubator atmosphere, but rather are concerned with changes in the PO$_2$ of the medium generated by the cells themselves and its effect on cell function. By far the most common problem is lack of adequate oxygen in cell culture. All mammalian cells require oxygen for their metabolism. This oxygen is obtained from the surrounding medium. In stationary cultures the oxygen enters the medium by diffusion from the medium–atmosphere interface. If the diffusion rate is too low, the cells are too far from the medium surface or the cells are using the oxygen too fast, and oxygen deprivation, poor cell growth, and eventually death will result. Oxygen generally will only diffuse freely through the first few millimeters from the surface. The amount of medium in a culture dish or flask will therefore be a balance between the cell's need for nutrients over the course of the experiment and the rate at which oxygen can diffuse from the surface to the cell monolayer. Too much medium will inhibit oxygen diffusion and slow cell growth; too little may be exhausted leading to cell death. Figure 3.4 demonstrates the effect of changing the depth of medium in the dish. While medium volume is different, the suspension-adapted cells are plated at the identical cell density/ml medium. Therefore, the major effect in these experiments is the limitation of oxygen

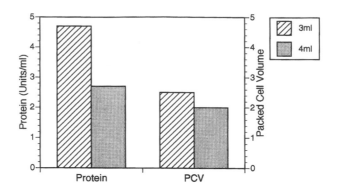

Figure 3.4. The effect of using different medium volume on cell number. Too much medium in a plate results in an increase in depth of the medium over the cells and a decreased oxygen content. This can frequently result in decreased cell growth and function (e.g., protein secretion).

diffusion to the cells in the dishes containing more medium. Further experiments carried out in fermenters with the PO_2 controlled to different set points confirmed that the growth of these cells was highly dependent on maintaining a proper oxygen level.

One constant threat to cell health is oxidative damage. Some cells are so sensitive to such damage that they cannot be maintained *in vitro* without antioxidants being added to the medium. One such cell type is the primary porcine Leydig cell, which will die between 72 and 90 hr of culture unless antioxidants such as vitamin E are added to the culture (Mather *et al.,* 1983). This extreme sensitivity to an oxidative environment is found only in Leydig cells from the pig, although functional alterations may be present in other cell types when antioxidants are not present. In contrast, some cells do better in a reducing environment created by the addition of β-mercaptoethanol or other reducing agents. It should be recognized that reducing agents (including mild reducing agents such as cysteine found in many media formulations) will inactivate some hormones (e.g., insulin), which are required for cell growth.

SURFACES AND CELL SHAPE

A clear and obvious difference between the *in vitro* and *in vivo* environment is the surface to which cells attach and the shape they acquire as a result. While it seems intuitively obvious that cells in the organism do not grow attached to either glass or plastic, it is only in the last decade or two that researchers have demonstrated the profound effects that cell shape and the surface to which cells are attached can have on cell growth and differentiated function (for example, see Fig. 3.5). This is due in part to the recent explosion in the discovery of the components that make up the extracellular matrix (ECM), which is part of basement membranes and connective tissue in all organs. Many of these components (e.g., collagens, laminin, and fibronectin) are complex mixtures of components present in Engelbreth-Holm-Swarm and (EHS) tumor matrix ("matrigel"), and culture dishes precoated with matrix proteins are now commercially available. Recent research has also expanded our understanding of the "receptors" that the cells use to attach to these ECM proteins and how this attachment affects cell function (Ashkenas *et al.,* 1996; Mather *et al.,* 1984). An

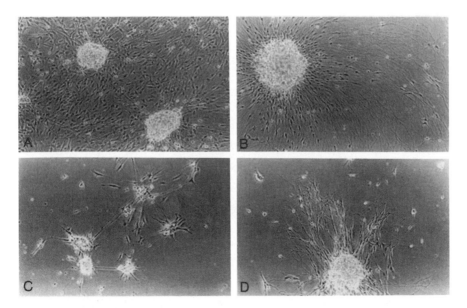

Figure 3.5. The effect of extracellular matrix on Schwann cell growth and survival. Primary cultures of rat dorsal root ganglia were grown for 2 days on (A) laminin-, (B) polylysine-, (C) fibronectin-, or (D) collagen-coated plates.

added complication is the fact that many cells *in vitro* can make their own attachment factors and can even secrete quite complex ECM (Gospodarowicz *et al.,* 1981; Mather and Phillips, 1984), thus altering the surface they are plated on to one more amenable to that cell's function. Figure 3.6 shows the complex matrix produced by the TR-1 testicular capillary endothelial cell line *in vitro.*

Serum also contains soluble attachment factors, such as fibronectin, which rapidly bind to tissue culture plastic and provide a modified surface for cell attachment. Cells may also add to or break down extracellular proteins that are provided for them. These ECM substrate proteins are discussed more in Chapter 8 on serum-free medium. We will review different types of surfaces that are available for growing cells and their uses below.

ADHERENT VERSUS NONADHERENT CELLS

Most cells in an animal grow attached to some structure in connective tissue, basement membrane, or mineral matrix such as bone. Cells in the blood, lymph, and other fluids are the only ones that normally grow "in suspension." Even so, many of these, such as lymphocytes, have the ability to move into solid tissue when required. Many of the cells which grow *in vivo* as attached cells cannot grow *in vitro* as a single cell suspension. If they are placed in culture in such a way that they cannot attach to a surface or another cell, they will die. Cells from tumors and those transformed *in vitro* frequently lose this need for attachment. In fact, the ability of cells to grow suspended in soft agar is considered one of the characteristics of transformed cells.

However, not all cells that grow in suspension are necessarily transformed. Lymphocytes readily grow in suspension and many cell types will survive and even function better when grown in conditions in which they do not attach and spread on a cell surface, but rather

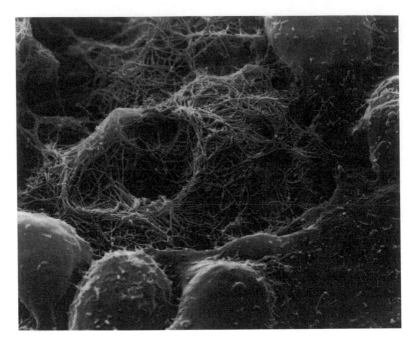

Figure 3.6. Extracellular matrix produced by a testicular capillary endothelial (TR-1) cell line. The scanning electron micrograph clearly shows the extensive matrix laid down by the cells *in vitro*. The cells are adhering to the matrix. The matrix consists of collagen fibers and amorphous materials, including laminin and fibronectin.

grow attached to each other as small aggregates. This highlights an important distinction between cell attachment and cell shape. Cells in a small cluster that are attached to each other are not growing as a single cell suspension but do have a more rounded shape than they would if they were attached to and spread on a surface. When interpreting the outcome of experimental manipulations, care must be taken to differentiate between the effects of changes in culture surface and the effects of changes in cell shape, which may result from altering the surface on which they are grown.

PLASTIC—DIFFERENT TYPES FOR DIFFERENT PURPOSES

The commercial availability of relatively cheap, sterile plasticware for tissue culture use has been crucial to the tremendous increase in the use of cell culture in the last 40 years. Now virtually all the pipettes, plates, filter units, and so forth needed for cell culture can be purchased as disposable sterile plastics. This not only has made maintaining the high-use laboratory less labor intensive, but it has made it feasible to have a laboratory where cell culture is only a small part of the work, since there is no necessity to invest in autoclaves, special equipment for washing culture glassware, and so forth. These savings can offset the higher expense of nonreusable cultureware. There is a bewildering variety of culture plasticware currently available. Terasaki plates contain only 10 μl of medium per well, while roller bottles can contain 100 ml and grow very large numbers of cells. Culture plastic comes with surfaces treated to have different charges or precoated with attachment factors or complex matrix. A selection of these culture dishes is shown in Fig. 2.11. A laboratory that performs many different types of culture may stock a dozen or more different types of culture dishes for specialized uses.

The type and configuration of culture vessel chosen will depend on the purpose of the culture. If producing a large number of cells (10^6–10^8) is the primary purpose of the culture, then larger plates such as 100 mm or 150 mm diameter may be optimal. Roller bottles may be used to grow even larger numbers of cells (10^8–10^9), but the cells are more difficult to remove and some percentage are lost. If producing large volumes (1–10,000 liter) of conditioned medium is the aim, then roller bottles, spinners, or fermenters will be the vessel of choice. Alternatively, for growth experiments, the multiwell plates are frequently an ideal choice, saving reagents and requiring fewer cells (high throughput growth assays are discussed in Chapter 11). However, if dishes are to be harvested at different times, individual 35- or 60-mm plates may be better. If the cells are to be observed or photographed, then very small wells present difficulties. Cells may instead be grown on coverslips or on chambered plastic or glass slides for use in immunocytochemistry, or for preparation for scanning or electron microscopy (see Chapter 6).

When switching back and forth between different dish configurations, it is important to maintain a constant inoculum density (as measured in cells/mm^2 of dish surface) and a similar amount of medium/cell. The latter may have to vary, however, since meniscus formation in small wells (e.g., 24-, 48-, or 96-well plate wells) may require adding proportionately more medium/well in order to keep the cells in the center submerged. Table 2.1 lists some commonly used dish sizes, their surface area, and the amount of medium commonly added per well.

GLASS

Originally, all cell and tissue culture was carried out in glass plates or vessels. This method is extremely labor intensive, since it requires that all plates be carefully washed and resterilized after use. The availability of cheap disposable sterile plasticware has made culture of cells on glass a special-purpose exception rather than the rule. In some cases it may be necessary to have cells grown on glass. The manipulation involved in fixing, staining, and viewing cells for imunohistochemistry, *in situ* mRNA analysis, or scanning electron microscopy, for example, are best carried out on cells attached to a glass slide or coverslip. There are several ways to do this. Sterile "chamber slides" containing 1, 2, 4, 8, or 16 chambers per slide are commercially available. This glass (or plastic) slide has a plastic superstructure attached by a silicon gasket that divides the slide into the requisite number of sections and contains the cells and medium. This is protected by a plastic cover. The cells are plated onto the glass surface in the usual medium and placed in the incubator. When the cells have reached their desired state of confluence, the slides are removed and the cells fixed and/or stained for observation. The same cells can be placed in all chambers and each well stained individually, or different cells can be placed in each well and the entire slide stained at one time. The plastic grid can be removed from the slide and the slide can be handled like a normal glass slide for staining, mounting, and observation using an upright microscope and short working distance objectives. This method has the advantage of flexibility and convenience; however, some cells will not stick well to these surfaces.

An alternate method is to grow cells directly on coverslips placed in a plastic tissue culture dish. Several square coverslips can be placed in a 60- or 100-mm plastic tissue culture dish or round coverslips can be individually placed into the wells of 24-well multiwell plates. The coverslips must be sterilized. It is best to clean them thoroughly to remove any chemicals and coatings left on by the manufacturer. This may be done by soaking a number of coverslips for 2 hr in acetone, rinsing and soaking overnight in ethanol, and after a second ethanol rinse flaming to sterilize (or let coverslips drain dry). Each can then be placed in a dish with a sterile forceps. The glass surface can be coated with attachment factors and the cells cultured as usual. The coverslips and attached cells can then be fixed and stained and mounted directly on a slide for viewing. It is easiest to handle No. 2 thickness coverslips, which do not break as easily as the thinner ones. However, these may be too thick for the optimal working distance of some objectives.

CELL SHAPE

The shape and size of cells *in vivo* will depend on their environment, what they are attached to, the hormonal signals they are receiving, and their function. Thus, epithelial cells may form a single cell layer attached to a basement membrane with thick, highly polarized cells. Peritubular cells may also be attached in a single layer but may be very thin, acting more as a physical barrier than a protein or mucopolysaccharide-secreting factory. Interstitial cells such as those in the ovary and testis may be almost round in cross-section and may occur in connected clumps in the tissue. In the artificial environment of *in vitro* culture systems, most cells will be rounded when grown in suspension and flattened when grown on a surface such as tissue culture plastic. Some cells, such as pancreatic β cells or testicular Leydig cells, may show increased function when they are allowed to maintain a rounded shape; other cell types may need to attach and spread to survive (Folkman and Moscona, 1978; Gospodarowicz *et al.,* 1978).

BASEMENT MEMBRANE AND ATTACHMENT FACTORS

Plastic or glass surfaces can be coated with a purified attachment protein such as fibronectin, laminin, or collagen. This is discussed more completely in Chapter 8 on serum-free culture. Basement membrane can also be laid down on the culture dish by one cell type, the cells removed sterilely, and a second cell type cultured on this cell-produced ECM (Gospodarowicz *et al.,* 1980, 1982; Mather *et al.,* 1984). This is less defined and more laborious than using purified protein components, but might well support a cell's growth or function better. A procedure for producing plates coated with cell-produced ECM is outlined later in this chapter. In addition, a cell can be selected to produce the tissue-type-specific ECM that would be most appropriate for the cell to be grown *in vitro,* thus providing a more physiologically relevant culture model. For example, a stromal cell line may be used to produce ECM to grow the epithelial cell type normally found adjacent to that stromal cell.

ARTIFICIAL MEMBRANES

Cells can also be grown on a variety of artificial membranes. Generally speaking, there are several advantages to growing cells on membranes. The membranes can be used to separate the cell from the medium or one cell from another, and thus study cell communication and cell secretion. Cells will extend processes into the pores of some membranes or even migrate through membranes if the pores are large enough. Membranes can be chosen to have a specific molecular weight cutoff, and thus can pass or retain nutrients or cell-secreted products on the basis of size. In this case, the cells must be contained on one side of a continuous membrane and the other side will contain a separate medium reservoir. Most

Table 3.1
Various Types and Configurations of Membranes that Can Be Used as Cell Substrates *in Vitro*

Membrane material	Porosity/pore size	Plate configuration	Optical properties	Comment
PET	1–14%/ 0.4–8.0 μm	6-, 12-, 24-well	Low pore density, high optical quality	Immunohistochemistry, transport
PTFE	80%/ 0.4–8.0	6-, 12-, 24-well	Clear when wet, cell outlines visible	High porosity for, maximum diffusion, requires matrix
Polyester	0.4–0.8 μm	6-, 12-, 24-well	Clear	TC treated
PC	0.4–8.0	6-, 12-, 24-well	Translucent, cells not visible	Usually requires some matrix

Abbreviations: PET, polyethylene terephthalate; PTFE, polytetrafluoroethylene; PC, polycarbonate; TC, tissue culture.

membranes are opaque, which makes direct observation of the cells growing on them impossible. However, several types of membranes have been developed for cell culture that are almost transparent when wet and allow observation of the cells in an inverted microscope. Cells grown on membranes can be fixed and processed for either scanning or electron microscopy if the membrane material is compatible with the solvents and fixatives used in these processes.

Membranes are also produced as hollow fibers within a sterile reservoir with inlets and outlets, as flat sheets or disks, or attached to plastic rings that can be inserted into a standard cell culture dish. While the use of membranes is expensive, one can obtain data in these configurations that cannot be obtained in other ways. Table 3.1 describes a variety of membranes used in membrane-containing inserts and hollow fiber culture systems. Figure 3.7 shows an electron micrograph of cells grown on a membrane and sectioned perpendicular to the membrane.

STRESS

Another role of the supporting environment is to protect the cells in culture from stress that might kill them directly, set off natural genetic programs that may lead to their death (apoptosis), or cause mutations or other changes that would not reflect their natural functioning *in vivo* or their desired function *in vitro*. Stress encountered by the cell *in vitro* can be due to rapid changes in pH, temperature, or osmolality, which set off the transcription of "heat-shock" genes in response to this stress. Agents that damage DNA such as mutagenic chemicals or even light may also lead to cell death or mutation. Agents that compromise membrane stability, such as detergentlike compounds (including high concentrations of vitamin A and steroids), may lead to leakiness and the inability of the cells to maintain the pH and ion gradients necessary for function. Oxidative damage to cells is also an ever-present threat, given that oxygen is a necessary part of the cultural environment. Cells may be differentially sensitive to such damage, depending on their cytoplasmic enzymes, cellular composition, and ability to produce protective enzymes such as superoxide dismutase. Finally, cells may be sensitive to contaminants and breakdown products found in medium, such as heavy metals contaminating medium components, organic compounds from insufficiently pure water, or ammonia produced as a by-product of cellular metabolism.

Figure 3.7. Transmission electron micrograph of a section through an RL-65 Clara cell line grown on a membrane. The cells can be seen above the filter and extending cellular processes down into the filter. The thickness of the filter is greater than that of the cells. Medium was added to the well beneath the filter while the upper surface of the cells was in contact with air.

Again, it is important to keep in mind the difference between the goal of re-creating an *in vitro* environment that allows the cells to function, to the greatest extent possible, the same way they do *in vivo* and other equally valid aims of culture. These model systems are invaluable for increasing our understanding of mammalian development, normal function, and disease states. However, there are many instances in which one might wish to use the flexibility and mutability of single cells *in vitro* to create a cell that is designed for a specific purpose, even if this might not reflect the *in vivo* situation. One such instance is the creation and optimization of secretion of monoclonal antibodies by hybridomas.

If one wishes to use cells to provide a model for studying *in vivo* physiology, the cells should be protected from stress as much as possible from their initiation as a cell line, and certainly during the experimental procedures. Serum supplementation protects against many of these factors, since it is designed to protect cells in the organism from similar stress by maintaining homeostasis. Serum has significant buffering capacity, proteins that bind

and detoxify heavy metals, and significant protease inhibitors to protect against damage from proteases released by living and lysed cells. Thus serum-free culture requires both more care and, to the greatest extent possible, replacement of protective properties by defined components such as organic buffers, protease inhibitors, and antioxidants. On the other hand, few cells exist in serum except in a wound. The presence of serum may therefore prime these cells to respond the way they would respond *in vivo* to this type of trauma, thus compromising the study of normal cell function.

If a cell line is to be engineered for a specific purpose, however, selecting for maximum resistance to stress may be desirable. This may be done by subjecting the cell line to sublethal stress and selecting for those cells most resistant. In fact, this has happened to most of the commonly used cell lines in the process of their being passed from laboratory to laboratory over the years.

pH, TEMPERATURE, AND OSMOLALITY

Generally, the best idea is to protect cells from rapid changes in pH, temperature, and osmolality. These may shock the cells and lead to apoptotic death. Many researchers prefer not to remove the cells from the incubator at all, or even open the incubator, while a crucial experiment is in progress. This is obviously not possible in most laboratories or desirable where direct observation of cultures may add significant data. Incubators that contain several inner doors to the chambers may minimize drift in temperature and pH when the door is opened. In any case, all users should take care to keep incubators open for the minimum amount of time. Organic buffers, as discussed above, can be used to minimize pH swings.

MECHANICAL

Mechanical stress is generally not a concern to those growing cells in stationary culture such as tissue culture dishes or flasks. In fact, mechanical stress is a normal part of the environment of many cell types, including heart myocytes and other muscle cells that are stretched regularly, cells in the bloodstream that undergo a good deal of shear when traveling through small vessels and capillaries, and cells lining the lung airways that are stretched with each breath. Some experimenters re-create these stresses *in vitro* to study their effects. For example, myocytes can be grown on flexible substrates that can be stretched to study the effects of mechanical stress on these heart cells. In most cases, however, mechanical stress is best minimized *in vitro*.

Roller bottle culture represents minimal mechanical stress, but spinners, fermenters, and other methods of large-scale culture all present a certain amount of mechanical stress and the potential for mechanical damage to cells. These stresses include a significant change in pressure in large (e.g., 10,000 liter) tanks as the cells move from the top to the bottom of the tank, shear created by the mixing propellers and the bubbles created during oxygenation of the tank, and shear created when the cells are transferred through pumps from one vessel to another. Large-scale culture vessels and media are designed to minimize this mechanical stress. See Chapters 11 and 12 for more discussion of growth of cells in roller bottles and large-scale cultures.

TOXIC CHEMICALS AND HEAVY METALS

Elements in the chemical environment of the cell such as heavy metals can also damage cells if they are not bound and sequestered. As mentioned above, many proteins in serum provide this protection, but serum can also contain toxic substances. Since these vary from serum to serum lot and the sensitivity varies from cell line to cell line, serum lots should be tested and compared using the cell lines you are planning to culture. In some cases, several passages in a new lot of serum may be required to expose a detrimental effect, but the tests outlined in Chapter 4 are generally adequate.

The best course to follow, in the absence or presence of serum, is to buy components of high purity, prepare tissue-culture-grade water in the laboratory for making media, and keep glassware and weighing devices for use exclusively in making media. These precautions are discussed in detail in Chapter 4. In serum-free medium, transport proteins such as transferrin and ceruloplasmin can act not only to provide needed nutrient to the cells but to protect against damage caused by free iron, copper, and other metals that may also bind to these proteins.

PROTEASES

Proteases can be secreted by viable cells or released from dying cells into the medium. These may directly injure the cells, remove cell surface components such as receptors that are necessary for proper cell growth, or degrade growth factors added to the medium or proteins secreted by the cells. Serum contains protease inhibitors and proteases and substances such as plasminogen, which can be converted into active proteases by cell-secreted plasminogen activator. Thus, the proteolysis in serum-containing medium will reflect the cell type and amount of serum present. If cells are grown serum free, nonspecific protein such as bovine serum albumin or specific protease inhibitors can be added to protect from proteolytic degradation. Protease inhibitors are discussed more fully in Chapter 8.

The best-controlled physical environment still needs the addition of a well-thought-out chemical and hormonal environment to produce the optimal growth and function of cells *in vitro*. The selection and preparation of the cell culture medium is discussed in Chapter 4 and the design of serum-free and defined culture media are discussed in Chapter 8.

REFERENCES

Ashkenas, J., Muschler, J., and Bissell, M. J., 1966, The extracellular matrix in epithelial biology, *Dev. Biol.* **180**:433–444.

Folkman, J., and Moscona, A., 1978, Role of cell shape in growth control, *Nature* **273**:345–349.

Gospodarowicz, D., Greenberg, G., and Birdwell, C. R., 1978, Determination of cellular shape by the extracellular matrix and its correlation with the control of cellular growth, *Cancer Res.* **38**:4155–4177.

Gospodarowicz, D., Delgado, D., and Vlodansky, I., 1980, Permissive effects of the extracellular matrix on cell proliferation *in vitro*, *Pro. Natl. Acad. Sci. USA* **77**:4094–4098.

Gospodarowicz, D., Greenberg, G., Foidart, J. M., and Savion, N., 1981, The production and localization of laminin in cultured vascular and corneal endothelial cells, *J. Cell Physiol.* **197**:171–183.

Gospodarowicz, D., Cohen, D., and Fujii, D. K., 1982, Regulation of cell growth by the basal lamina and plasma factors: Relevance to embryonic control of cell proliferation and differentiation. in Sato, G. H., Pardee, A. B., and Sirbasku, D. A., *Growth of Cells in Hormonally Defined Medium.* Cold Spring Harbor Laboratory, Cold Spring Harbor, NY, pp. 95–124.

Mather, J. P., and Phillips, D. M., 1984, Establishment of a peritubular myoid-like cell line and interactions between established testicular cell lines in culture, *J. Ultrastruct. Res.* **87**:263–274.

Mather, J. P., Saez, J. M., Dray, F., and Haour, F., 1983, Vitamin E prolongs survival and function of porcine Leydig cells in culture, *Acta Endocrinol.* **102**:470–475.

Mather, J. P., Wolpe, S. D., Gunsalus, G. L., Bardin, C. W., and Phillips, D. M., 1984, Effect of purified and cell-produced extracellular matrix components on Sertoli cell function. *Ann. NY Acad. Sci.* **438**:572–575.

Waymouth, C., 1970, Osmolality of mammalian blood and of media for culture of mammalian cells, *In Vitro* **6**:109–127.

CHAPTER 4

Media

In this section, we will deal with the nutrient mixtures that are usually called *media*. In actual use, these will almost always be supplemented with serum or another complex biological fluid (milk, embryo extracts, and plasma are examples), or, as discussed in Chapter 8, with a defined mixture of hormones and growth factors. The ongoing experimental work of replacing the complex mixtures with defined components, both nutrients and proteins, largely has been responsible both for our understanding of what the medium does in cell culture and for our increased technical ability to maintain a broad range of functional cells *in vitro*.

All media are composed primarily of water. If media are prepared in the laboratory, the water source can make a critical difference in the quality of the media prepared. This is especially important for serum-free media where trace metals and organic compounds present in some distilled or deionized water can cause severe toxicity problems for some types of cells. Water quality is discussed in more detail in Chapter 8.

————— WHAT DOES THE MEDIUM DO? —————

The nutrient mixture is the cornerstone of cell culture. Having the correct nutrient mixture can often be the determining factor in failure or success in growing a cell *in vitro*. Almost all the other components of the cell culture system can be substituted for by simpler methods. For example, one can put cells and medium in a flask, blow into the flask through a plugged pipette to bring the CO_2 to 5%, cap the flask, and sit on it to maintain the appropriate "incubation temperature" (while this is tedious, it works!). The nutrient mixtures can be substituted for only by using even more complex fluids, which sometimes fail to support cell growth. It has been determined that mRNA levels for some cellular proteins will decrease after as little as 20 min in a simple glucose–salt-buffered solution.

The medium provides essential nutrients that are incorporated into dividing cells, such as amino acids, fatty acids, sugars, ions, trace elements, vitamins, and cofactors, and ions and molecules necessary to maintain the proper chemical environment for the cell. Some components may perform both roles; for example, the sodium bicarbonate may be used as a carbonate source but also may play an important role in maintaining the appropriate pH and osmolality. The medium contains all or part of the buffering system required to maintain a physiological pH (see below on incubator settings) and should provide the appropriate osmolality of the cells. Nutrients include amino acids, with the richer media containing both "essential" and "nonessential" amino acids. Media also contain lipids; most contain a mixture of fatty acids, and some contain more complex lipids (e.g., cholesterol). Some media formulations, such as medium 199, contain detergents (e.g., Tween 80) to help emulsify the lipids. These detergents can be toxic to some types of cells, particularly in serum-free medium. Some media contain macromolecules such as thymidine, adenosine, and hypoxanthine that can be synthesized by cells *in vitro*. Adding more of these to the medium may nonetheless improve the growth of some cells by maintaining an appropriate pool size of precursors in the cells. Many media contain the common vitamins such as niacin, folic acid, riboflavin, inositol, thiamine, and so forth (see Table 4.1). While these vitamins are essential to continued cell replication, a detrimental effect may not be seen until several cell doublings after their removal from the medium. Other vitamins such as vitamins D (1,25-dihydroxycholecalciferol), C (ascorbic acid), E (α-tocopherol), and A (retinol, retinoic acid) are not commonly added to media formulations because they are unstable in solution. (Even if these are in the medium formula, they may not be active by the time the medium gets to the cells.) However, these may prove beneficial or even essential for some cell types and should be added separately. They may also be involved in maintaining the differentiated state of the cell, in regulating cell functions, or acting as antioxidants.

All media contain some energy source, usually glucose, although the molar levels can vary widely (0.8–>5 g/liter). Amino acids and glucose, as well as ions such as NaCl, contribute to the osmolality of the medium, as well as having a nutritional role. In addition to the bicarbonate–CO_2 buffering system, the medium may also contain some phosphate buffer and perhaps complex organic buffers. Medium may also contain antioxidants or reducing agents (or these might be added separately). Most media contain phenol red as a pH indicator. This is very helpful in rapidly assessing the pH of the medium of all the cultures in an incubator. Phenol red can be added to media if it is not part of the medium powder or if a more obvious color is desired. It should be noted than phenol red has weak estrogenic activity, which may be a consideration with some cells.

The composition of several commonly used media are compared in Table 4.1. Most media [e.g., minimal essential (ME) medium, Dulbecco's Modified Eagle's (DME) medium] were developed specifically for use with serum supplementation and high density growth of cells (Dulbecco and Freeman, 1959; Eagle, 1955). In contrast, others such as Ham's nutrient mixtures F12, F10, and the Molecular Cellular Developmental Biology (MCDB) series of media were tailored specifically for growing a given cell type at low density with a minimal amount of undefined protein added, so as to study the effects of the nutrient components of the media (Ham, 1965; Ham and McKeehan, 1979). The F12–DME (1:1, v/v) medium was originally devised for growing cells in defined serum-free conditions (Mather and Sato, 1979) (now commercially available as a premixed powder). F12–DME medium works well for growing cells at low or high densities and in defined conditions or with serum. Picking the optimal medium for a specific use will be discussed below. Note in Table 4.1 that the media for growing mammalian cells have similar components, while the insect cell culture media (e.g., Grace's medium) is quite different, reflecting the different metabolic needs of insect cells. Leibovitz L-15 medium is designed to grow cells in equilibrium with air rather than CO_2–air and is useful when CO_2 incubators are not available (e.g., the teaching laboratory), or when cells are shipped or handled extensively outside the incubator (for example, during a long tissue dissociation protocol).

Table 4.1
Comparison of Media Compositions[a]

						Media					
Components	BME	F10	F12	DMEM	F12–DME	Med 199	RPMI 1640	Waymouth	L15	CMRL 1	Grace's Insect Media
Inorganic salts											
Calcium chloride (CaCl$_2$)	200	33.29	33.22	200	116.6	200		90.6	140	200	750
Calcium nitrate							100				
Cupric sulfate (CuSO$_4$–5H$_2$O)		0.0025	0.002		0.001						
Ferric nitrate (Fe(NO$_3$)$_3$–9H$_2$O)				0.1	0.05	0.7					
Ferrous sulfate (FeSO$_4$–7H$_2$O)		0.834	0.834		0.417						
Potassium chloride (KCl)	400	285	223.6	400	311.8	400	400	150	400	400	4100
Potassium phosphate, mono.		83						80	60		
Magnesium chloride (MgCl$_2$)	97.7				28.64			112.6	93.7		2280
Magnesium sulfate (MgSO$_4$)		74.62	57.22	97.67	48.84	97.67	48.84	100	97.67	97.7	2780
Sodium bicarbonate	2200	1200	1176	3700	2438	2200	2000	2240		2200	350
Sodium chloride (NaCl)	6800	7400	7599	6400	6995.5	6800	6000	6000	8000	6800	
Sodium phosphate, mono.	140	153.7	142	125	62.5	140			190	140	1013
Sodium phosphate, dibas.					71.02		800	300			
Zinc sulfate (ZrSO$_4$–7H$_2$O)		0.0288	0.863		0.432						
Other compounds											
Adenine sulfate						10					
Adenosine 5'-triphosphate						1					
Adenosine 5'-phosphate						0.2					
Cholesterol						0.2				0.2	
Cocarboxylase										1	
Coenzyme A										2.5	
D+Galactose									900		
D+Fructose											400
2'Deoxyadenosine										10	
2'Deoxycytidine										10	
2'Deoxyguanosine										10	
Deoxyribose						0.5					
Diphosphopyridine nucleotide (NAD)										7	

(continued)

Table 4.1 (*Continued*)

Components		BME	F10	F12	DMEM	F12–DME	Med 199	RPMI 1640	Waymouth	L15	CMRL 1	Grace's Insect Media
	Media											
Flavin adenine dinucleotide (FAD)											1	
Fumaric acid												55
D-Glucose		1000	1100	1802	4500	3151	1000	2000	5000		1000	700
Glutathione (reduced)							0.05	1	15		10	
Guanine hydrochloride							0.3					
Hypoxanthine-Na			4.7	4.77		2.39	0.4		29			
Alpha-ketoglutaric acid												370
Lactalbumin hydrolysate												3330
Linoleic acid				0.084		0.042						
Lipoic acid			0.2	0.21		0.105						
Malic acid												670
5-Methyl-deoxycytidine											0.1	
Phenol red		10	1.2	1.2	15	8.1	20	5	10	10	20	
Putrescine-2HCl				0.161		0.081						
Ribose							0.5					
Sodium acetate							50					
Sodium acetate-3H₂O											83	
Sodium glucuronate-H₂O											4.2	
Sodium pyruvate						55				550		
Succinic acid			110	110								
Sucrose											60	26680
Thymine							0.3					
Thymidine			0.7	0.7		0.365					10	
Triphosphopyridine nucleotide (NADP)											1	
Tween 80							20				5	
Uracil							0.3					
Uridine 5'-triphosphate											1	
Xanthine-Na							0.3					
Yeastolate												3330

Component	1	2	3	4	5	6	7	8	9	10	11
Amino acids											
Beta-Alanine	200										
L-Alanine	225	25	225			25	4.45			9	21
L-Arginine hydrochlor	700	70	500	75	200	70	147.5	84	8.9	211	211
L-Asparagine-H_2O	350		250				7.5		15	15	
L-Aspartic acid	350	30		60	20	30	6.65		35.1	13.3	13
L-Cysteine-HCl-H_2O		260	120	100	65	0.1	17.56			25	16
L-Cystine-2HCl	22	26		20	20	75	31.29	63			
L-Glutamic acid	600	75		150	20	75	7.35		14.7	14.7	16
L-Glutamine	600		300	350	300	100	365	584	146		292
Glycine	650	50	200	50	10	50	18.75	30	7.5	7.5	
L-Histidine-HCl-H_2O	2500	20	250	164	15	21.88	31.48	42	21	23	10.81
Hydroxyproline		10			20	10					
L-Isoleucine	50	20	250	25	50	40	54.47	105	4	2.6	26
L-Leucine	75	60	125	50	50	60	59.05	105	13.1	13	26
L-Lysine hydrochloride	625	70	75	240	40	70	91.25	146	36.5	29	36.47
DL-Methionine											
L-Methionine	50	15	75	50		15	17.24	30	4.5	4.5	7.5
DL-Phenylalanine											
L-Phenylalanine	150	25	125	50	15	25	35.48		5	5	16.5
L-Proline	350	40				40	17.25	66	34.5	11.5	
DL-Serine	550										
L-Serine		25	200	50	30	25	26.25	42	10.5	10.5	10.5
DL-Threonine			300								
L-Threonine	175	30		75	20	30	53.45	95	11.9	3.6	24
L-Tryptophan	100	10	20	40	5	10	9.02	16	2.04	0.6	4
L-Tyrosine-2Na-2H_2O	50	58	300	58	29	58	55.79	104	7.81	2.62	26
L-Valine	100	25	100	65	20	25	52.85	94	11.7	3.5	23.5
Vitamins											
Ascorbic acid		50		17.5		0.05					
Alpha-tocopherol						0.01					
Biotin	0.01	0.01		.02	0.2	0.01	0.0035		0.0073	0.024	
Calciferol (vitamin D2)						0.1					
d-Calcium pantothenate	0.02	0.01	1	1	0.25	0.01	2.24	4	0.5	0.7	
Choline chloride	0.2	0.5	1	250	3	0.5	8.98	4	14	0.7	

(continued)

Table 4.1 (*Continued*)

Components	Media										
	BME	F10	F12	DMEM	F12–DME	Med 199	RPMI 1640	Waymouth	L15	CMRL 1	Grace's Insect Media
Folic acid	1	1.3	1.3	4	2.65	0.01	1	.4	1	0.01	0.02
L-Inositol	2	0.5	18	7.2	12.6	0.05	35	1	2	0.05	0.02
Menadione (vitamin K3)						0.01					
Niacin						0.025				0.025	0.02
Niacinamide	1	0.6	0.036	4	2.02	0.025	1	1	1	0.025	0.02
Para-aminobenzoic acid						0.05	1			0.05	
Pyridoxal hydrochloride	1					0.025				0.025	
Pyridoxine HCl		0.2	0.06	4	2	0.025	1	1	1	0.025	0.02
Riboflavin	0.1	0.4	0.037	0.4	0.219	0.01	0.2	1		0.01	0.02
Riboflavin 5'-phosphate, Na									0.1		
Thiamine hydrochloride	1	1	0.3	4	2.17	0.01	1	10		0.01	0.02
Thiamine monophosphate									1		
Vitamin A (acetate)						0.14					
Vitamin B12		1.4	1.4		0.68		0.005	.2			

[a]This table compares the composition and concentrations of components of several media. Note the qualitative and quantitative differences in the media.

47

MATCHING
THE
INCUBATOR
SETTINGS
AND THE
MEDIUM

Table 4.2
Commonly Used Media[a]

Media	Applicable to	Reference
Basal media Eagle (BME)	Growing cells with serum	Eagle (1965)
Minimal essential media (MEM)	Growing cells with dialysed serum	Eagle (1959)
Dulbecco's Modified Eagle's media (DMEM)	Many virus transfected cells, growth with serum, high-density growth	Dulbecco and Freeman (1959)
Ham's F10 media		Ham (1963)
Ham's F12 nutrient mixture (F12)	CHO cells, low density, low serum protein	Ham (1965)
F12–DME: (1:1) mixture	Serum growth, many cells, serum free	Mather and Sato (1979); Bottenstein et al. (1979)
William's media E	Rat liver epithelial cells	Williams and Gunn (1974)
RPMI 1630	Mouse leukemia cells, cells in suspension	
RPMI 1640	Human leukemic (and other) cells	Moore and Kitamura (1968)
Leibovitz L-15 medium	Buffered for air, human tumors	Leibovitz (1963)
Waymouth's MB 752/1	L cells	Waymouth (1959)
Fischer's media	Murine leukemia cells	Fischer and Sartorelli (1964)
McCoy's 5A media	Human lymphocytes	McCoy et al. (1959)
MCDB 131	Human endothelial cells	Knedler and Ham (1987)
Media 199	Chick embryo fibroblasts	Morgan et al. (1950)
Medium NCTC-109	Hybridomas	Evans et al. (1956)
Media NCTC-135	Serum, serum-free growth	Evans et al. (1964)
Neurobasal medium	CNS neurons	Brewer et al. (1994)
BGJb medium	Fetal rat long bones	Biggers et al. (1961)
Glasgow minimal media	BHK-21	Macpherson and Stoker (1962)
CMRL media	L-cells, monkey kidney cells	Parker et al. (1957)

[a]This table lists some of the commonly used media, the use for which they were originally derived, and the original reference. It is apparent that many of these media are used much more widely than originally intended. This is one reason that the time spent testing several media for optimal growth of a different cell type is time well spent.

More recently, vendors are supplying "special-use media" to grow a stated cell under special conditions. These sometimes contain undisclosed hormones, growth factors, or undefined protein components. Therefore, these cannot be considered "defined," although they may work well for some applications. Other such media are supplied with a defined supplement mix that must be added before use. A list of some commonly used media and the uses for which they were originally developed can be found in Table 4.2. Some special-use media and their suppliers are listed in Table 4.3. Tables 4.2 and 4.3 may be used as a rough guide to what media may be useful for what cells; however, many of these media are now used to grow a wide variety of cells. The optimization of the medium for specific cells and uses is discussed below.

MATCHING THE INCUBATOR SETTINGS AND THE MEDIUM

The CO_2 setting on incubators should be chosen to match the medium to be used. Each medium has been formulated with components designed to work with a specified CO_2 concentration (most ranging from 0 to 10% CO_2–air mixtures) to provide a bicarbonate buffer-

Table 4.3
List of Commercially Available Media Supplement
Mixtures and Their Contents

ITS (100×)
  Insulin–transferrin–selenium
Chemically defined lipid concentrate (100×)
  Arachadonic acid
  Cholesterol
  DL-α-tocopherol-acetate
  Linoleic acid
  Linolenic acid
  Myristic acid
  Oleic acid
  Palmitoleic acid
  Palmitic acid
  Pluronic F-68
  Stearic acid
  Tween-80
N-2 supplement (100×) neuronal cells
  Insulin
  Transferrin (human)
  Progesterone
  Putrascine
  Selenite
B-27 Supplement (50×) for fetal hippocampal neurons
MEM amino acids solution (100×)
BME vitamin solution (100×)
  NaCl
  Biotin
  D-Calcium pantothenate
  Choline chloride
  Folic acid
  L-inositol
  Nicotinamide
  Pyridoxal HCl
  Riboflavin
  Thiamine Hcl
MEM vitamin solution (100×)
  NaCl
  D-Calcium pantothenate
  Choline chloride
  Folic acid
  L-inositol
  Nicotinamide
  Pyridoxal HCl
  Riboflavin
  Thiamine HCl

ing system giving a pH of 7.0–7.4. Mismatch of medium bicarbonate levels and CO_2 incubator levels will result in the medium pH being out of the optimal range for cell growth, resulting in slower growth or cell death. If media designed for use with different CO_2 levels are to be used in the same incubator, the bicarbonate levels should be adjusted so that they all buffer correctly at the CO_2 level to which the incubator is set. It should be pointed out that the lowest CO_2 levels (with low bicarbonate) give a medium with a lower buffering capacity than a high CO_2–high bicarbonate system.

For example, our incubators are set at 5% CO_2. This is the correct setting to use with Ham's F12 nutrient mixture. However, this requires that the F12–DME medium be prepared with 1.2 g/liter $NaHCO_3$ (rather than the 2.4 g/liter that is the intermediate level for F12–DME). On the other hand, commercial DME medium is formulated for use with a 10% CO_2 environment and should have its bicarbonate level reduced for use with 5% CO_2. The ability to alter the bicarbonate concentration when the medium is made is clearly one advantage of preparing liquid medium in the laboratory from commercial powders. Considerations that should be kept in mind when altering the CO_2–bicarbonate levels are that the bicarbonate concentration itself is important for some cell types; that altering the bicarbonate changes the total buffering capacity of the medium, as discussed above; and that altering the bicarbonate changes the osmolarity of the medium.

—— HOW TO SELECT THE APPROPRIATE MEDIUM ——

If a new cell line is brought into the laboratory, determine what medium is recommended for its growth. This information can be obtained from the same source as the cells. If the recommended medium is incompatible with the CO_2 settings on the incubator used for other cells grown in the laboratory, or is not commonly prepared in the laboratory, one might wish to change the growth medium. It is best to initially grow the cells in their original medium and compare this with the more convenient medium after a passage or two in each. If the growth rate and morphology of the cells look the same, then a medium switch can be made. However, keep in mind when trying to repeat published data that cells grown in a different medium may respond differently in some parameters measured. For example, estrogen-requiring cells previously grown in F12 medium and then switched to DME medium may have a diminished response to exogenously added estrogens in this medium because of the weak estrogenic activity of the phenol red, which is present at a much higher concentration in the DME medium.

If one wishes to grow a primary cell in culture and no published data exist on growing that cell type *in vitro,* or one wishes to grow the cells in a different manner (e.g., with defined supplements rather than serum), it is best to screen several of the commercially available media before deciding on the one that is best for that particular use. This can be done by obtaining five to ten candidate media powders from a supplier, preparing them all in the laboratory as described below using the same water and supplementary components, and doing a direct comparison of cell growth in the different conditions (see Fig. 4.1). If endpoints other than cell growth are important, measure these too in each of the media. Carry the cells in the selected medium for several passages and freeze them in this medium for future use.

It is interesting to compare the formulations of the media tested and look for patterns of whether the cells of interest seem to prefer a low or high glucose, low or high calcium, low or high levels of amino acids, and so forth. This will be useful if you decide to further modify the medium, as described in the section below. It is important to know the composition of the test medium if conditioned medium is to be screened for biological activity. Since this is often done at 10–50% conditioned medium, the test medium should minimally use the same bicarbonate buffer concentration as the assay medium and optimally be identical to the assay medium except for changes introduced by the conditioning cells.

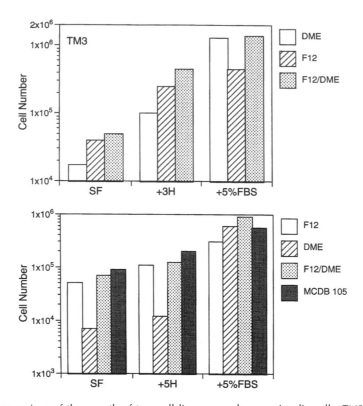

Figure 4.1. Comparison of the growth of two cell lines: normal mouse Leydig cells, TM3, and mouse melanoma cells, M2R, in several different media using different media supplements (SF, serum free, no supplements; 3H or 5H, the defined supplements appropriate for that cell type; or 5% of FBS). Note that the cell number is presented on a log scale. The choice of optimal medium will depend on the cell line and the medium supplements to be used.

MEDIA PREPARATION

Media can be purchased as prepared liquid media, made up in the laboratory from dried powders containing most of the components of the nutrient mixtures, or prepared in the laboratory from individual stocks of the individual components or groups of the components. We do not recommend purchasing liquid medium, especially for serum-free culture work. Medium components deteriorate with time, and do so faster in solution. Some necessary components break down and are lost, others create toxic breakdown products or oxidize to toxic components. While it varies from cell to cell and with serum-supplemented or serum-free media, we have found that 2 weeks is a safe storage time for serum-free media, or longer if serum is added when the medium is prepared. Outdated medium can be used for washing cells or preparing tissues for primary culture. Clearly, this is not adequate time to allow for commercial preparation and storage, shipping, and further storage of medium in your laboratory. Some prepared liquid media can be frozen. Those that form a precipitate when thawed should not be frozen. In any case, it is always safe to store the prepared powdered medium and make liquid medium in the laboratory on a regular basis.

Powdered nutrient mixtures generally have a shelf life of a year or more if stored in moisture-proof, airtight containers in the dark. If the laboratory does not use large volumes of media, the 1-liter packages are convenient. Preparing medium in the laboratory from components immediately before use is obviously the best way to insure that the medium contains the desired components in the desired form. This is essential if the investigator wishes to study the role of the nutrients themselves or to optimize the nutrient portion of the medium as described below (Ham and McKeehan, 1979). However, most laboratories will find that preparation of medium from commercial powdered nutrient mixtures and a limited storage of the prepared media in a light-tight refrigerator will be adequate for their needs. This is also less costly than purchasing prepared media, especially when the cost of filters and so forth can be spread over large-volume use.

As mentioned above, F12–DME medium can support the growth of a large variety of mammalian cells, although other basal media have been shown to be optimal for specific cell types (Fig. 4.1). We will use F12–DME medium as the medium of choice for most of the exercises in this volume. We prepare F12–DME medium using the commercial powder formulation and add 1.2 g/liter $NaHCO_3$, 0.4 g/liter glutamine, and if the medium is to be used for serum-free culture, 15 mM HEPES buffer. We use low glucose DME medium when mixing the two powders. The addition of 1.2 g/liter bicarbonate provides the correct pH when used in a 5% CO_2 incubator. HEPES provides additional buffering capacity in the cultures and stabilizes the pH during the time that the cultures are out of the incubator and at normal atmosphere for observation and manipulation. Serum itself has considerable buffering capacity; so, if serum (5–15%) is to be used as a supplement, the HEPES concentration can be reduced to 10 mM or eliminated. We do not add antibiotics to the medium as prepared below. This insures that poor sterile technique will be rapidly detected, as discussed in Chapter 7. All antibiotic agents have some toxicity, so any antibiotic to be used should be tested over a relevant concentration range on the cells of interest. When preparing primary cultures, an antibiotic may be added to the wash medium during the initial stages of tissue handling. This may be done by preparing a 1000-fold concentrated stock solution of an antibiotic such as gentamycin (in phosphate-buffered saline) and adding it directly to the wash medium. If HEPES is used, a 1-M stock solution is prepared for this purpose.

Generally, wide-mouthed glass Schott bottles are used for storing medium. The maximum shelf life is 2 weeks at 60°C for serum-free medium and 1 month for serum-containing medium.

PREPARING MEDIUM

Reagents given are for preparing 1 liter of medium.

Materials

1. DME powder 6.6 g (no bicarbonate, low glucose)
2. F12 powder 5.3 g (no bicarbonate); or premixed F12–DME powder, 11.9 g (no bicarbonate)
3. $NaHCO_3$, 1.2 g
4. Glutamine, 0.4 g
5. 1 M HEPES buffer

52

MEDIA

Procedure

1. Fill an Erlenmeyer flask with 750 ml purified water.
2. Add DME powder slowly to aid in dissolving and avoid clumping.
3. When DME is in solution, add F12 powder.
4. When F12 is in solution, add $NaHCO_3$.
5. Add glutamine.
6. QS to final volume with purified water (See Glossary for definition of QS).
7. pH the medium (7.2).
8. Remove 15 ml of the medium.
9. Add 15 ml of a 1 M HEPES buffer.
10. Stir for an additional 10 min.

FILTER STERILIZATION

Materials

1-liter bottletop filter unit, 0.22 μm filter (Corning) for Schott bottles, or 1-liter filter flasks. If more than 1 liter at a time is routinely made, the following apparatus may be more practical for filtration:
1. Variable flow peristaltic pump, 1 liter/min
2. Pump head (polycarbonate, with stainless steel rotor), to accommodate tubing
3. Thick-wall silicon tubing (4.8 mm inner diameter)
4. Millipore 20-liter Millipak filters

Procedure

Clamp the tubing in the pump head and attach the pump head, tightening just enough to keep it in place. Adjust the tubing so that there is sufficient length in the Erlenmeyer reservoir and sufficient length in the outbound side of the pump to reach the receiving bottles at the filter end. Tighten the pump head. Attach the tubing to the filter (clamping is usually not necessary). Adjust the flow rate slowly at first and vent the filter by loosening the filter vent cap until there are no air bubbles in the filter unit. The flow rate can now be increased to maximum, but care should be taken that any back pressure does not force the tubing off the filter's hose barb.

The smallest filter that will handle the volume of medium to be sterilized without plugging should be used. Different lots of serum can vary considerably in their filterability. Horse serum and adult bovine serum have a higher lipid and protein content than fetal bovine serum and can be more difficult to filter. A prefilter can be used to extend the filter life and increase filtration rate. A much larger volume of serum-free medium than serum-containing medium can be filtered through a filter of a given area.

Serum can be added to medium when it is made up and the mixture filtered. However, since the serum has a longer shelf life than the medium and different cells may require different levels or kinds of serum, we find it more convenient to make up the medium without serum, store the sterile serum separately, and add it to the medium as needed for each experiment. Whether using disposable filters or a filtration setup, the first 30 to 100 ml of medium through the filter should be discarded to avoid contaminating the medium with chemicals washed out of the filter. The media bottles should be labeled, dated, and stored at 4°C.

SERUM TREATMENT

53

TESTING
MEDIA
AND
COMPONENTS—
QUALITY
CONTROL

Serum is frequently spoken of as if it were a defined single substance. This is very far from the truth. Cell culture media can be supplemented with sera from any species of animal; bovine (fetal, newborn, or adult), horse, or human sera are the most frequently used. These are quite different in many ways and can have very different effects on the properties of cells grown in them. Additionally, serum varies from animal to animal, with changes in diet, and seasonally. Therefore, there is considerable variability from lot to lot of the commercially available sera. In addition to whole sera, which is allowed to clot and the clot removed, the blood can be collected with an anticlotting agent and the cellular portion spun out, resulting in plasma. Serum and plasma, even from the same animal, are quite different in composition and their effect on cells.

Sera can be treated before use in one or more ways: filtration, dialysis, diafiltration, heat treatment, or fractionation. These treatments can act as an added insurance against contamination, can remove or inactivate toxic components of the serum, can remove or inactivate growth-promoting or differentiating components of the serum, and specifically can remove low- or high-molecular-weight components of the serum or particular serum fractions. Clearly, this complex and undefined addition to medium must be treated with some care to insure consistent results. The only way to insure good results is to thoroughly test several lots of serum for their ability to support the desired cell characteristics (e.g., growth, differentiation, or lack of differentiation; specific biochemical markers; protein production, etc.) and then buy a large quantity of the best lot, store it at −20 to −80°C, and use it for the next several years.

Most commercial sera come sterilely packaged. It is best to purchase serum that has been sterilely collected as well, as an added insurance against viruses or mycoplasma, which can go through some filters. When adding human serum to cultures, the entire culture and all waste should be treated as a biohazard. Human sera should be collected from known donors or blood banks that test for the common viruses such as human immunodeficiency virus (HIV) and hepatitis.

TESTING MEDIA AND COMPONENTS—QUALITY CONTROL

We have stated that it is best to prepare media from commercially available powdered nutrient mixtures as outlined above. It is important to keep good records and do quality control testing of reagents used in making the medium. We generally keep one set of glassware exclusively for medium making. This is rinsed well with distilled water but not washed with detergent between each use. This avoids the possibility of any detergent residue getting into the medium. The sodium bicarbonate and other reagents that are used in media should be used only for media. When weighing out these reagents, a disposable tongue depressor and weigh boat should be used. This avoids contaminating these reagents with other, potentially toxic chemicals that may be in use in the laboratory.

The major component of the medium is water. Water purity is very important for good-quality medium (see Chapter 2 for stills and filtration devices). Again, water quality can be more critical when cells are grown in serum-free medium. However, some cell types can be extremely sensitive to poor medium quality even when serum is used. Figure 4.2 shows growth curves for two different cell lines in media that differ only in the water used to make

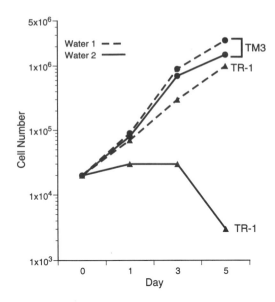

Figure 4.2. The effect of water quality on the growth of two different cell lines in defined medium. The TR-1 endothelial cell line is much more sensitive than the TM3 line. Dashed lines, Milli-Q purified water; dotted lines, "house"-distilled water. Water quality should always be the best possible, since this can be an extremely difficult problem to track down.

them up. It is obvious that the TR-1 cells (a capillary endothelial-derived cell line) cannot be grown at all in serum-free medium made with poor-quality water, although they will grow at a decreased rate if this medium is supplemented with serum. Studies such as that of Mather *et al.* (1986) have determined that heavy metals and organic compounds can account for some of the toxicity in poor-quality water.

We like to lot test media powders, critical hormones, and sera and buy enough of the best lot to last a year. Lot testing would include testing the previously used lot of medium, for example, against two or more new lots. All three batches of medium should be made up on the same day, with the only difference being the medium lot (e.g., same water, sodium bicarbonate, serum, etc.). Testing should include daily growth curves from which one can calculate lag, population doubling time, and saturation density (see Chapter 5). It is also a good idea to do a plating efficiency experiment, since cells at low density are more sensitive to toxic components, while high-density growth would be better to detect a medium deficient in nutrients.

——— TROUBLESHOOTING MEDIUM PROBLEMS ———

Even when these precautions are followed, there will come a time when problems arise that require troubleshooting. This is when meticulous testing and record keeping pay off. Follow the steps below to identify and eliminate the problem:

1. Talk to all persons using the culture facility. Determine whether the problem is being experienced in many different cell lines or only a few; by all users or only a few.
2. When did the problem start? Determine the earliest date that anyone thought they might have a problem.

3. Are there any reagents that are used only with cells having a problem; or by all cells having the problem?

4. Were new lots of any of the medium or supplemental reagents put into use at or within 1 or 2 weeks before the problem started?

5. Were new lots of tissue culture plates or a different brand used about this time?

6. Test all cell lines in the laboratory for mycoplasma and other potential contaminants. If they are all contaminant free and only one cell line seems to be having trouble, thaw out a vial of cells from an earlier freeze of that line.

7. If any of the above questions have turned up a suspicious reagent or supply, test this first. Make up a medium using a different lot of medium powder, serum, etc. (or open another lot of tissue culture dishes) and test comparing the newly made medium to the presumptive "bad" lot of medium. It is best to change only one thing at a time.

8. Get water from another source (e.g., a still in another laboratory) and test medium made with this water.

While all of the precautions outlined above may seem excessive, good quality control can save days and weeks spent tracking down problems that affect experimental outcome and can make the difference between success and failure in growing some types of cells. We have experienced many problems over the years, including seasonal variation in distilled water quality; serum lots that will support the growth of one cell type in the laboratory, but not others; serum whose inadequacy to support growth was only apparent after 4–5 passages; plasticware to which cells would not attach (Fig. 4.3); a medium powder lot miss-

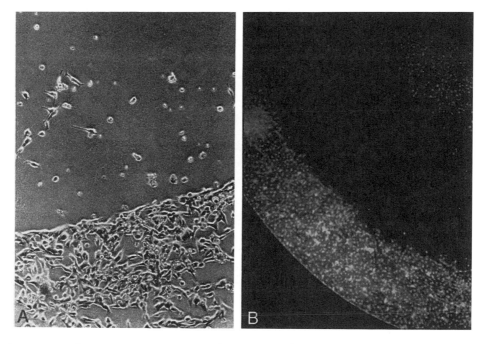

Figure 4.3. An illustration of poor cell growth due to poor quality tissue culture plates at (A) high and (B) low magnification. Note that the cells are not attaching evenly to the wells but grow well in some portions and poorly in others. A similar bull's-eye pattern of growth can be seen if cells attach slowly in an incubator that vibrates excessively.

ing one component; and many more. Even the best-run laboratory will inevitably experience problems. If you do trace the problems experienced in your laboratory to a specific reagent or lot of culture dishes, notify the manufacturer. They will usually be helpful in correcting the problem and/or replacing the defective materials.

_____ **ALTERING COMMERCIAL MEDIA** _____
FOR SPECIAL USES

Sometimes an addition to a commercial medium can much improve cell growth. If cells are being grown at high densities and are very lactogenic (turn the medium acidic rapidly), the addition of more glucose to the medium may improve growth and prolong viability. One can try adding a trace element mix, such as those described by Ham _et al._ (Hamilton and Ham, 1977; McKeehan _et al.,_ 1976, 1977). Many of these trace elements are normally provided as trace contaminants that enter the medium in water or serum. As the medium becomes more defined and the water more pure, these trace elements need to be added to the medium formulation. Sometimes an increased concentration of vitamins can be useful. Vitamin mixtures are commercially available and can be added as such. Some vitamins such as vitamin E (or α-tocopherol) or vitamin A (retinol or retinoic acid) are not added to media mixtures because of their instability, but they might be important for some cells to survive or function _in vitro._

An example of this is shown in Fig. 4.4. These primary cultures of porcine Leydig cells would all die on the third day of culture unless vitamin E was added to the medium. This effect could be mimicked by other antioxidants such as vitamin C (Mather _et al.,_ 1983). While few cultures show this dramatic a dependence on antioxidants, many types of cells

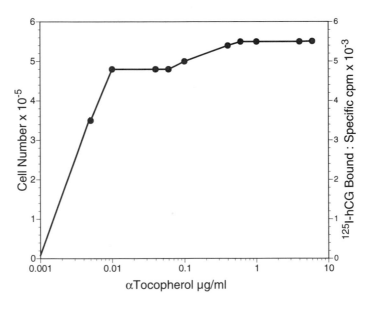

Figure 4.4. The effect of vitamin E (alpha-tocopherol acetate) on the survival of porcine Leydig cells in primary culture. Vitamin E is essential if the cells are to survive beyond the third day of culture.

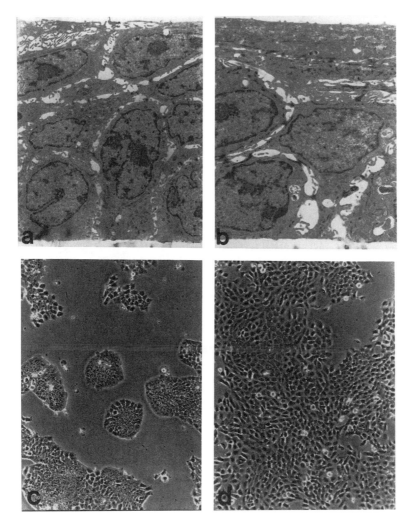

Figure 4.5. The effect of the addition of vitamin A on the phenotype of the RL-65 cell line (C). The vitamin A-supplemented cells are more tightly packed and appear smaller in phase. As can be seen in the electron micrographs, however, the (A) vitamin A treated cells are less keratinized and more columnar in shape than the (B,D) nontreated controls, and thus reflect a more normal state for this lung cell type.

benefit by having an antioxidant included in the medium. These should be made up and stored as a concentrated stock solution and added to the medium immediately before the addition of cells (see Chapter 8).

Another example of the usefulness of supplementing media with vitamins is shown in Fig. 4.5. These are RL-65 cells, a cell line derived from Clara cells in the lung (Roberts *et al.*, 1990). It was noticed that these cells became increasingly difficult to remove from the dish at subculture when they were left for more than a few days between subcultures. The cells seemed very resistant to trypsinization, and many cells were lost on subculture. On closer examination, it was apparent that some of the cells were differentiating into a keratinized layer on top of the monolayer. This layer of keratinized cells excluded the trypsin

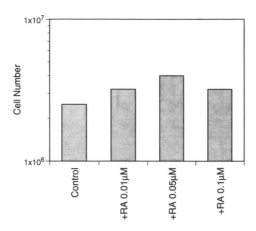

Figure 4.6. The effect of vitamin A on cell number of RL-65 cells. Cells such as those shown in Fig. 4.5B,C were counted. Even though the cells in Fig. 4.5C look more sparse than those in 4.5D, more cells are present.

and were, of course, dead. The addition of retinoic acid to these cultures prevented keratinization and helped promote more reliable growth (Fig. 4.6). Interestingly, this ability of the RL-65 cell line reflects an *in vivo* property of the airway epithelium, which can keratinize in states of extreme vitamin A deficiency. A different effect of retinoic acid *in vitro* is demonstrated by the primary rat Sertoli cell cultures. These cells are postmitotic, and thus there is no effect of retinoic acid on growth. However, as shown in Fig. 4.7, retinoic acid was a major factor in maintaining transferrin secretion by these cells *in vitro* (Perez-Infante *et al.,* 1986).

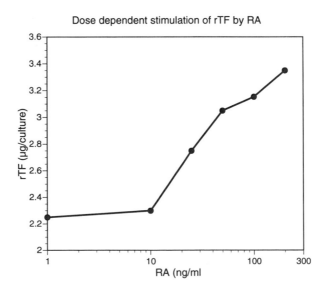

Figure 4.7. The effect of vitamin A on transferrin secretion by Sertoli cells in primary culture. In this case the same vitamin A stimulates the secretion of a specific protein in cultures of nondividing cells.

MEDIUM OPTIMIZATION

Many scientists have devoted their careers to understanding the role of nutrient mixtures in supporting cell growth and survival *in vitro*. These studies have resulted in the nutrient mixtures currently published or commercially available. There is, however, still a need for more experimentation to derive the optimal media for other cell types (e.g., newly derived cell lines, human cell lines, etc.) or other culture needs (e.g., very-high-density culture, controlling differentiation through culture conditions). Optimizing the medium in which a primary culture or cell line is grown can lead to increased growth, increased protein secretion, increased viability, increased phenotypic stability, and better control of differentiation. Optimizing the nutrient mixture is an important part of this process.

The best way to optimize the nutrient mixture is to sequentially perform dose–response curves on each component, select the optimal range for each, and retest each component. This must be done as an iterative process because the ratios of the components, as well as the absolute levels, are critical in optimizing the medium. This process should be done using the desired endpoint to screen. For example, if a medium is to be optimized to achieve maximal recombinant protein secretion, then the screen should be done using protein titer as the endpoint assayed. If growth is to be optimized, then cell number is the endpoint. Medium optimized for one parameter will not necessarily be best for others. An idealized example of three different dose–response curves from a screen to optimize media for protein secretion by CHO is shown in Fig. 4.8. Note that the cells will tolerate a broad range of concentrations for some components, but will have a very narrow optimal concentration range for others. It may also be possible to substitute some components for others, for ex-

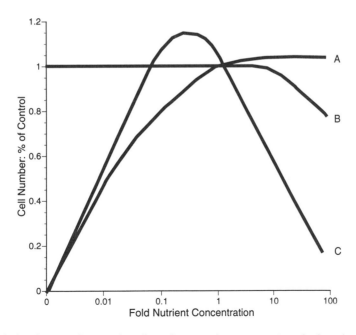

Figure 4.8. Idealized curves showing the effect of varying the concentration of selected nutrients on cell growth. (A) A required nutrient with a broad optimum (e.g., leucine). (B) A required nutrient with a tight optimal range (e.g., KCl). (C) A nutrient with little effect when added alone.

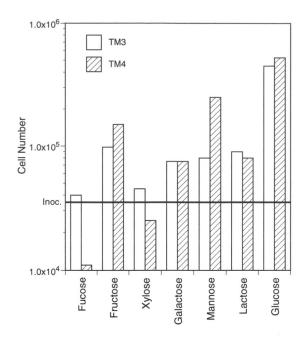

Figure 4.9. The effect of substituting different sugars for glucose on the growth of two cell lines TM3 (testicular Leydig-derived) and TM4 (testicular Sertoli-derived).

ample, substituting a different energy source for glucose (Fig. 4.9) or combining energy sources.

All these tests should be done in the presence of the medium supplement that will be used (e.g., serum, growth factor mix, etc.). For optimizing the hormone and growth factor additions for serum-free culture, see Chapter 8. Most investigators will not wish to go to the expense and time required to optimize specifically for their function. In this case, commercial media may be used, but it is wise to spend a minimum of time determining which of the available options is best. The steps for this "quick and dirty" optimization technique follow:

1. Obtain and make up medium from several different nutrient mixtures. Be sure and use the appropriate bicarbonate level for your incubator settings (see above).
2. Supplement media with the required supplements (serum or hormones). If serum reduction is a desired goal, run a dose–response curve for serum and choose a serum concentration that gives 50% of the optimal growth. This will allow one to detect any "serum-sparing" effects of the media.
3. Choose the best of the media tested.
4. Try adding additional glucose, especially if the cells are to be grown to high density.
5. Try supplementing this medium with various sterile supplement solutions (see Table 4.3 for commercially available mixtures) such as additional glutamine, vitamins, amino acids, nonessential amino acids, and so forth.
6. Try supplementing the medium with insulin and transferrin (5 μg/ml, each).
7. Try varying the osmolarity up or down.

8. Thoroughly compare cells grown for one and for four or five passages in the original medium and the newly selected medium to see how several different parameters compare.

9. Freeze down cells grown in old medium (in case there are long-term changes that you do not want) and grow cells in the new medium. Freeze more cells down after three to four passages in the medium.

REFERENCES

Biggers, J. D., Gwatrikin, R. B. L., and Heynes, S., 1961, Growth of embryonic avian and mammalian tibiae on a relatively simple chemically defined medium, *Exp. Cell Res.* **25:**41–58.

Bottenstein, J., Hayashi, I., Hutchings, S. H., Masui, H., Mather, J., McClure, D. B., Okasa, S., Rizzino, A., Sato, G., Serrero, G., Wolfe, R., and Wu, R., 1979, The growth of cells in serum free hormone supplemented media, *Methods Enzymol.* **58:**94–109.

Brewer, G., Torricelli, J., Evege, E., and Price, P. J., 1994, Neurobasal super (TM) medium/B27 supplement: A new serum-free medium combination for survival of neurons, *Focus* **16:**6–8.

Dulbecco, R., and Freeman, G., 1959, Plaque Production by the Polyoma Virus, *Virology* **8:**396–405.

Eagle, H., 1955, Nutrition needs of mammalian cells in tissue culture, *Science* **122:**501–504.

Eagle, H., 1959, Amino acid metabolism in mammalian cell cultures, *Science* **130:**432–437.

Eagle, H., 1965, Propagation in a fluid medium of a human epidermoid carcinoma strain KB, *Proc. Exp. Biol. Med.* **89:**362–364.

Evans, V S., Bryant, J. C., Kerr, H., and Schilling, E., 1964, Chemically defined media for cultivation of long-term cell strains from mammalian species, *Exp. Cell Res.* **36:**439–448.

Evans, V. S., Bryant, J. C., Fioramonti, M. C., McQuilkin, W. T., Sanford, K. K., and Earle, W. R., 1956, Studies of nutrient media for tissue cells *in vitro*. I. A protein free chemically defined medium for cultivation of strain Leydig cells, *Cancer Res.* **16:**77–86.

Fischer, G. A., and Sartorelli, A. C., 1964, *Development, Maintenance and Assay of Drug Resistance,* in: Eisen, H. N. (ed.), *Methods in Medical Research.* Vol. 10, Year Book Medical Publishers, Chicago, pp. 247–262.

Ham, R., 1965, Clonal growth of mammalian cells in a chemically defined synthetic medium, *Proc. Natl. Acad. Sci. USA* **53:**288–293.

Ham, R. G., 1963, An improved nutrient solution for diploid Chinese hamster and human cell lines, *Exp. Cell Res.* **29:**515–526.

Ham, R. G., and McKeehan, W. L. (eds.), 1979, Media and growth requirements, *Methods Enzymol.* **58:**44–93.

Hamilton, W., and Ham, R., 1977, Clonal growth of Chinese hamster cell lines in protein-free media, *In Vitro* **13:**537–547.

Knedler, A., and Ham, R. G., 1987, Optimized medium for clonal growth of human microvascular endothelial cells with minimal serum. *In Vitro Cell. Dev. Biol.* **23:**481–491.

Leibovitz, A., 1963, The growth and maintenance of tissue-cell cultures in free gas exchange with the atmosphere, *Am. J. Hyg.* **78:**173–180.

Macpherson, I. A., and Stoker, M. 1962, Polyoma transformation of hamster cell clones—An investigation of genetic factors affecting cell competence, *Virology* **16:**147–151.

Mather, J., Kaczarowski, F., Gabler, R., and Wilkins, F., 1986, Effects of water purity and additional of common water contaminants on the growth of cells in serum-free media, *BioTechniques* **4:**56–63.

Mather, J. P., Saez, J. M., Dray, F., and Haour, F., 1983, Vitamin E prolongs survival and function of porcine Leydig cells in culture, *Acta Endocrinol.* **102:**470–475.

Mather, J. P., and Sato, G. H., 1979, The growth of mouse melanoma cells in hormone-supplemented, serum-free medium, *Exp. Cell Res.* **120:**191–200.

McCoy, T. A., Maxwell, M., and Kruse, P. F., 1959, Amino acid requirements of the Novikoff hepatoma *in vitro, Proc. Soc. Exp. Biol. Med.* **100:**115–118.

McKeehan, W., Hamilton, W., and Ham, R., 1976, Selenium is an essential trace nutrient for growth of WI-38 diploid human fibroblasts, *Proc. Natl. Acad. Sci. USA* **73:**2023–2027.

McKeehan, W. L., McKeehan, K. A., Hammond, S. L., and Ham, R. G., 1977. Improved medium for clonal growth of human diploid fibroblasts at low concentrations of serum protein. *In Vitro* **13:**399–416.

Moore, G. E., and Kitamura, H., 1968, Cell line derived from patient with myeloma, *NY State J. Med.* **68:**2054–2060.

Morgan, J. F., Morton, H. J., and Parker, R. C., 1950, Nutrition of animal cells in tissue culture. I. Initial studies on a synthetic medium, *Proc. Soc. Exp. Biol. Med.* **73:**1–8.

Parker, R., Castor, L. N., and McCulloch, E. A., 1957, Altered cell strains in continuous culture, *NY Acad. Sci.* **5:**303–313.

Perez-Infante, V., Bardin, C. W., Gunsalus, G. L., Musto, N. A., Rich, K. A., and Mather, J. P., 1986, Differential regulation of testicular transferrin and androgen-binding protein secretion in primary cultures of rat Sertoli cells, *Endocrinology* **118:**383–392.

Roberts, P. E., Phillips, D. M., and Mather, J. M., 1990, Properties of a novel epithelial cell from immature rat lung: Establishment and maintenance of the differentiated phenotype, *Am. J. Physiol. Lung Cell. Mol. Physiol.* **3:**415–425.

Waymouth, C., 1959, Rapid proliferation of sublines of NCTC Clone 929 (Strain L) mouse cells in a simple chemically defined medium (MB 752/1), *J. Natl. Cancer Inst.* **22:**1003–1015.

Williams, G. M., and Gunn, J. M., 1974, Long-term cell culture of adult rat liver epithelial cells, *Exp. Cell Res.* **89:**139–142.

CHAPTER 5

Standard Cell Culture Techniques

This chapter will describe the relatively few standard techniques that are the basis for the majority of manipulations of cells *in vitro*. Once one masters these few techniques, he or she will be able to perform routine maintenance of cell lines and set up cells for most biochemical or cell culture experiments. The more complicated techniques needed for special types of culture, the special considerations to be kept in mind when doing serum-free culture, and preparation of cells for primary culture and large-scale culture are all dealt with in separate chapters. The simplicity of these basic methods makes cell culture an ideal tool in teaching laboratories and for students. A minimum of time can be spent on techniques, allowing more time to spend on teaching the scientific method and allowing the students to discover the biology of organisms through cell cultures. Do not be misled, however; the simplicity of the techniques is not indicative of a simple discipline. It may take only 2 weeks to learn how to grow cells when everything goes right, but it can take 20 years to accumulate the knowledge to know what to do when things go wrong! Since the literature of cell culture is a long and rich one spanning many years, much of this information is not easily accessible. We will try and impart not only technical direction, but a basic understanding of the biology behind these techniques.

SUBCULTURING

Subculturing, or "splitting cells," is required to periodically provide fresh nutrients and growing space for continuously growing cell lines. The frequency of subculture and the split ratio, or density of cells plated, will depend on the characteristics of each cell line being carried. If cells are split too frequently or at too low a density, the line may be lost. If cells are not split frequently enough, the cells may exhaust the medium and die, or a different type of cell may be selected for in mixed cell cultures. One can even get significant phenotypic changes in cloned cultures that have the normal genetic drift expected with a normal spontaneous mutation rate and a doubling time of 24 hr or less. In general, once the correct timing and split ratio are found for a particular cell line, it should be used consistently for that line, with only minor variations when absolutely essential. Different subculture strategies can select for different properties in the cell lines carried. Varying the subculture routine may lead to a state of confusion on the part of the cells—and the scientist.

Subculture involves removing the growth media, washing the plate, disassociating the adhered cells, usually enzymatically (e.g., with trypsin, although some cells may be removed by repeated pipetting or gentle scraping), and diluting the cell suspension into fresh media. If this involves the use of serum and the split ratio is high (e.g., 1:100), it is usually not necessary to remove the residual enzyme. If the culture is maintained in serum-free media, however, it is necessary to neutralize the enzyme by using an appropriate protease inhibitor, such as soybean trypsin inhibitor. There are other considerations that require that different methods be used. For example, if the cell line is carried at a very low split ratio (e.g., 1:2–1:5), it is advisable to wash the cells after enzyme treatment even if serum is present. These special cases will be described in the appropriate sections. All solutions added to dishes/flasks with adherent cells should be added by pipetting along the side of the dish or flask to avoid disrupting the monolayer.

SUBCULTURING ADHERENT CELLS

Materials (all materials/solutions are sterile unless otherwise noted)

1. Growth medium
2. Phosphate-buffered saline (PBS) (Ca^{2+}/Mg^{2+} free)
3. Trypsin solution (0.05% (w/w) tissue-culture-grade trypsin, 0.53 mM EDTA; unless otherwise noted, this is the concentration used)
4. Culture dishes (100 mm)
5. Hemacytometer or electronic particle counter

Procedure

1. Remove medium from the plate. If the cell line adheres tightly, the medium may be discarded. If many cells in the plate are floating or only loosely attached, the plate should be gently shaken or washed with a pipette and the loose cells saved and recombined with the trypsin-dispersed cells before replating. If this is not done, the overall phenotype of the culture will change with time, as each passage will preferentially select for the more tightly adherent cells.
2. Wash 1× with 5 ml of PBS.
3. Add 2–3 ml trypsin. Allow the trypsin to cover the plate. Tilt plate and remove excess trypsin.
4. Incubate at 37°C. The time will vary depending on the cell type and whether it is a primary culture or an established cell line, but generally it will take 2–3 min. It is a good idea to remove the plate from the incubator and look to see if the cells have rounded up after about 2 min and check every minute thereafter. If the cells slough off the plate when its side is gently tapped against the bench, they are ready. Do not trypsinize beyond the time required to detach cells to this degree, since this will damage the cells and may reduce plating efficiency.
5. When the cells have rounded up and are coming off the plate, resuspend in 5 ml of serum-containing medium and wash cells by centrifugation at 800 rpm. Resuspend in 5 ml

medium. [This wash step may be omitted if the cells are to be split at a high split ratio (>1:50) in serum containing medium.] If using serum-free media, the trypsin should be neutralized with 1 ml of a 1 mg/ml solution of soybean trypsin inhibitor (STI), diluted to 10 ml with medium, and centrifuged in a clinical centrifuge for 3-4 min at 900 rpm. You must wash the cells if culturing serum free. This wash removes any residual enzyme and removes the STI, which can prevent attachment of some cells. The supernatant is then aspirated and the pellet resuspended by repeated pipetting in 5 ml of growth medium.

6. If the cells are primary or secondary cultures or the cell line is one with which you have had no experience, a high seeding density is recommended, i.e., 2.5 ml of cell suspension to 7.5 ml of growth medium (a 1:4 split ratio). If precise cell counts are needed or the required seeding density is known, an aliquot of the cell suspension should be counted at this time. It is a good idea to seed replicate stock plates at several densities, for example, 1×, 1/2×, 1/5×, and 1/10× the density chosen (or in the example above, 1:4, 1:8, 1:20, and 1:40 split ratio).

SUBCULTURING SUSPENSION CULTURES

Cultures of cells that grow in suspension in flasks or spinners can be maintained by diluting an aliquot of the suspension into fresh growth medium. See Chapter 11 for information on adapting attached cells to suspension culture and growing cells in spinners.

Materials (all materials are sterile)

1. Flasks (75 cm^2)
2. Growth medium
3. Hemacytometer or electronic particle counter

Procedure

1. Hold the flask upright and pipette the cell suspension up and down two or three times to disperse any clumps.
2. Either remove an aliquot for counting or, if precise counts are not required, transfer 200 μl to 1 ml of the suspension to a fresh flask containing 10 ml of growth medium. If a split ratio is less than 1:10, the appropriate volume of cell suspension should be placed in a 15 ml conical tube, diluted to 10 ml with medium, and centrifuged at 900 rpm for 3–4 min. Resuspend the resulting cell pellet into fresh growth medium and aliquot the appropriate number of cells into the number of flasks needed. In this way, one can avoid diluting fresh medium with exhausted medium components or carrying over toxic cell metabolites or proteases. *Note:* While many published methods suggest washing or diluting cells in PBS, we prefer to use serum-free medium (outdated medium can be used here). This seems to improve viability, especially for cells carried continuously in serum-free medium and when handling "delicate" or "picky" cells. Using serum to wash cells also prevents any rapid changes in osmolarity that may occur when switching from medium to PBS and back and maintains an energy source for the cells. Basically, the cells are less stressed during an already difficult period.

GROWTH CURVES AND MEASURING CELL GROWTH

In order to analyze the growth characteristics of a particular cell type or cell line, a growth curve can be established from which one can obtain a population doubling time, a lag time, and a saturation density. A growth curve generally will show the cell population's *lag phase,* that is, the time it takes for the cell to recover from subculture, attach, and spread; the *log phase,* in which the cell number begins to increase exponentially; and a *plateau phase,* in which the culture becomes confluent and the growth rate slows or stops (Fig. 5.1). An increase in cell number is also a frequently used method of assessing the effect of hormones, nutrients, and so forth on a specific cell type. *Growth,* or increase in total cell number over time, is a good measure of a biological response because it is so broadly defined and influenced by many different factors, including mitogens, changes in nutrient level, transport, membrane integrity, attachment factors, and so forth. However, merely seeing a difference in cell number or some secondary endpoint such as [^{3}H]thymidine incorporation, dye uptake, and so on gives only a limited amount of information. A factor can affect cell number by changing attachment, shortening or lengthening the lag phase, changing the plating efficiency (or survival at subculture), changing the death rate, changing the rate of progression through the cell cycle, or changing the plateau density. A growth curve will help differentiate between these effects.

While growth curves provide much information, they are too time and labor consuming to be used where changes in cell number are to be used as a screen. This may be the case if one wishes to assay the different fractions generated during the biochemical purification of a growth factor. Optimizing the nutritional components of a growth medium or screening conditioned media for growth-promoting activities also may be best performed using a

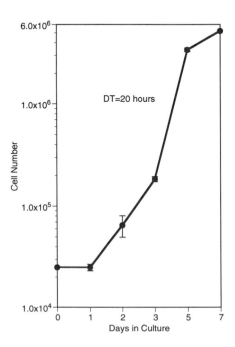

Figure 5.1. A typical growth curve of attachment-dependent cells. Note the lag phase before the cells start to grow, the log phase of growth, and the beginning of the plateau phase when contact-inhibited cells will cease growing. The population doubling time should be calculated during log phase growth and is 20 hr for this cell line (RL-65).

secondary endpoint for growth. Several of these will be discussed below in this section and more extensively in the section on high-throughput assays in Chapter 11. In addition to the changes in growth curves mentioned above, cells in different growth conditions can become multinucleate, increase or decrease total cell protein and volume, or undergo large changes in the levels of some mRNAs or enzymes. The secondary endpoint thus can vary to a different extent than the change in cell number or even in a different direction. Thus, if a secondary endpoint is to be used as a measure of cell proliferation, the assay used should be correlated with changes in cell number in both the control and test conditions.

USING THE HEMACYTOMETER OR ELECTRONIC PARTICLE COUNTER

Use of a hemacytometer or an electronic particle counter gives the most direct measurement of cell number and therefore cell growth. Cells can be counted before, during, and after setting up an experiment to accurately and directly quantitate and standardize experimental conditions. Moreover, the use of a dye such as trypan blue when doing hemacytometer counts gives the investigator a quantitative standard for the viability of the cells by doing a differential count of the cells that exclude trypan blue ("sort-of-sick" to viable) and those that take up the dye (irretrievably dead) (see also section on acridine orange–ethidium bromide assay). The hemacytometer is undoubtedly the cheapest and most labor intensive method for counting cells, but it can be used to provide data as accurate as that obtained by any other method and to provide an assessment of both total and viable cell counts. This makes it an ideal method for the student laboratory or for laboratories where cell counts are not performed frequently.

Materials

1. Trypsinized cell cultures
2. Improved Neubauer hemacytometer with coverslip
3. Tally counter
4. 0.4% trypan blue in PBS
5. Pasteur pipettes
6. Microscope

Procedure

1. Make a 1:1 dilution of cell suspension with 0.4% trypan blue. This can be further diluted with PBS if necessary.
2. Carefully resuspend with a Pasteur pipette.
3. Cover the hemacytometer chamber with the coverslip and place a drop of the suspension from the Pasteur pipette at the edge of the "V" shape on the chamber. Repeat for the other side of the chamber. It is important not to overfill or underfill, but rather to allow the drop to be drawn over the surface by capillary action.
4. Place the chamber on the stage of the microscope.
5. Initially focus on the etched lines of the chamber with low (4×) power (Fig. 5.2).
6. The hemacytometer consists of nine 1-mm squares that are divided into 25 smaller squares. The volume of one 1-mm square is 0.1 mm^3 or 10^{-4} ml. Using a 10× objective, focus on one of the 25 smaller squares bounded on all sides by three parallel lines.

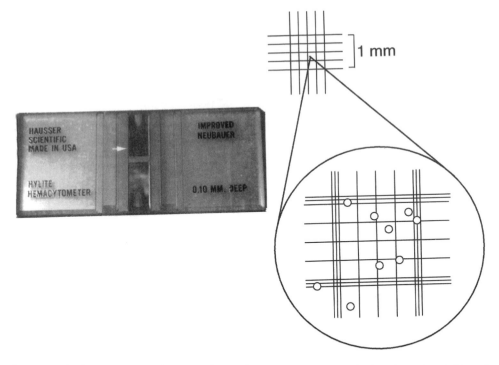

Figure 5.2. The hemacytometer shown can be used as an inexpensive method to determine viable cell number.

For a viable cell number, count all the cells that exclude the dye. To total cell number, viable cell number, and percent viability, separately count the blue cells and those that exclude the dye. A multichannel push-button counter is very useful here. In order not to count cells that lie on the border (the three parallel lines more than once), make a point of counting those cells that lie on the top and the right and not on the bottom and left (this really depends on your orientation: Do it any way that is easy for you, just do not count the same cell twice).

7. Count at least 100 cells/mm^2. If you count fewer than 100 cells in the square, count one or more additional squares.
8. Repeat for the second side of the chamber.
9. Calculate the number of cells/ml by multiplying the number of cells counted in 1-mm square (or the average of however many squares you counted) by 10^4.

ELECTRONIC PARTICLE COUNTING

The Coulter counter, manufactured by Coulter Electronics, affords the investigator a rapid, accurate, and reproducible method of total cell counting, particularly when dealing routinely with a large number of samples to count. Several models are currently available that enjoy widespread popularity among cell culturists (Fig. 5.3). As a cell passes through the aperture, through which an electrical current is flowing, it displaces an equal volume of electrolyte. This causes a change of resistance in the path of the current and subsequently a change in the voltage. This change in voltage is directly proportional to the volume or size of the cell. Every change in voltage during the sample flow is represented as a cell

Figure 5.3. The Coulter particle counter can be used for cell number determination. With this model counter, data collection and coincidence correction are automatically stored in the computer at right. The data can be easily transferred to a spread sheet for plotting.

count, which is then displayed on the LED readout. The counts shown include both viable and nonviable but intact cells. The counting threshold can be set to avoid counting cell debris. Cells must be a single-cell suspension to obtain accurate cell counts with these machines. The machines will also underestimate cell number in a manner related to the cell density. As the density becomes greater, there is an increasing chance that two cells will pass through the aperture so close together that they will only be counted as a single cell. This error can be corrected for by a mathematical formula that is used to generate the "coincidence correction tables" provided with the machines. This can be used to correct the counts manually in older model Coulter counters and is done automatically in the newer model counters.

Materials

1. Trypsinized cell suspension
2. Counting cuvette (blood dilution vials, Scientific Products)
3. Isoton solution

Procedure

1. Allow the counter to warm up prior to trypsinization.
2. Trypsinize cells as above.
3. Resuspend the cells in a total of 1 ml PBS and transfer to a cuvette. If this is a routine count for subculturing, remove 200 μl from the neutralized cell suspension and add it to a cuvette.

4. QS to 10 ml with Isoton solution.
5. Place the cuvette on the aperture platform and carefully immerse the aperture tube into the cuvette, taking care that the electrode is also immersed.
6. Open the stopcock (turn 90° to the vertical) on the aperture tube until the display resets to zero.
7. Close the valve.
8. The counter will sample 0.5 ml of the cuvette suspension.
9. Reset the counter and take a second count. This is important because in older machines particularly the polarity can affect the count itself; so, for complete accuracy, one should make two counts/cuvette.
10. If the count is greater than 10,000, check the number against the coincidence correction chart supplied by Coulter for this purpose. The machine is quite accurate between 10,000 and 100,000 cells/0.5 ml. If the count exceeds 100,000, dilute the cell suspension 1:2 or 1:4 and recount in the accurate range.
11. Calculate the cell number by multiplying the average of the two displays by 20 (volume of 10 ml), which gives the total number in the cuvette. This will also give the total number in the 1-ml sample. In the case where 200 μl is the sample volume, multiply by 5 to get the number of cells/ml. A shortcut when using 200 μl as a sample volume is to add two zeros to the readout on the display to get a direct number of cells/ml; that is, 51,234 will be read as 5,123,400 or 5.1×10^6 cells/ml. (*Note:* This number must be corrected for coincidence for an accurate count.) This is only accurate when a count of greater than 10,000 is displayed. If the cells had to be further diluted to get within the range of the machine, multiply the calculated cell number by this factor. *Note:* Mastering the calculations involved in determining total cell number from counter or hemocytometer raw counts is frequently the most difficult task for a beginning student to learn.

If the equipment is in working order and being used correctly, all means of obtaining total cell number, including the hemacytometer, Coulter Counter, and the more sophisticated FACS machines, should be in good agreement. As always, each has different potential sources of error. Not having a single cell suspension is the most frequent source of error for cell counts. Allowing the suspended cells to sit in PBS or another buffer for prolonged periods can also lead to cell death and, with fragile cells, the fragmentation of dead cells and an underestimate of the number of cells.

GENERATING A GROWTH CURVE

Materials

1. Cell cultures
2. Ten 60-mm tissue culture plates
3. Growth medium, 55 ml
4. Trypsin–EDTA solution
5. Soybean trypsin inhibitor (1 mg/ml stock), if serum free

Procedure

1. Trypsinize cells as previously described.
2. Resuspend cells in a total of 10 ml medium; this can be serum-containing medium, or if serum free, add 1 ml of soybean trypsin inhibitor.

3. Wash by centrifugation in a clinical centrifuge, 900 rpm 3–4 min.
4. Resuspend the pellet in 5 ml of medium; remove an aliquot for counting.
5. Dilute an appropriate amount of the cell suspension into 55 ml of growth medium so that the total cell number in the tube is 5.5×10^5.
6. Resuspend the cells well and add 5 ml to each plate. This should result in a seeding density of 5×10^4 cells/plate (2×10^{-3} cells/cm^2).
7. Count the remaining 5 ml of cell suspension to establish the actual seeding density. *Note:* It is wise to always count an aliquot of cells at the beginning and end of setting up any experiment. The counts should be identical. If these vary by more than 10%, there is a technical problem that should be corrected. If the variance is too great, the experiment should be discarded, since all plates will not have the same inoculum density.
8. Put the plates in a 37°C humidified incubator.
9. Count duplicate plates every 24 hr.
10. After the last day of counting, plot the results on a log-linear scale (Fig. 5.1). Determine the population-doubling time by identifying a cell number along the exponential phase of the curve, moving along the linear part of the curve until that cell number is doubled, and then calculating the number of hours between the two events.

SECONDARY ENDPOINT ASSAYS FOR PROLIFERATION

Sometimes it is preferable to use smaller plates or a more rapid readout method for assaying the proliferative effect of a factor or factors. This format becomes convenient when the factor to be assayed is in short supply. Alternatively, obtaining a sufficient number of cells may be a rate-limiting step. Often, the factor to be assayed needs to be titrated over a wide range or a large number of samples are to be tested. In such cases, a rapid readout format can be quite useful. This type of assay is also easier to adapt to high-throughput, including automated (i.e., robot), formats. Several methods for high-throughput assays are given in Chapter 11.

The three assays described here are a fluorescence-based assay, a colorimetric type, and the use of [^{3}H]thymidine incorporation. Each requires a different, fairly expensive piece of equipment to detect the readout. Equipment availability then might determine the assay of choice for your laboratory. Since each of these depends on a different aspect of cell physiology as its secondary endpoint, each is subject to different artifactual errors and technical difficulties. These are just three of many different alternative assays for measuring secondary endpoints that may (or may not) correlate with cell number. Other possible assays include total protein/culture, total DNA/culture, nuclear counts, lactate dehydrogenase or alkaline phosphatase enzyme measurement, packed cell volume, and so forth.

In the fluorescent assay, a probe, calcein-acetoxymethyl (AM) (see Appendix 5 supplier list for molecular probes) permeates the intact cell membrane. This is possible because the AM ester neutralizes the charged groups on the molecule to make it cell permeant. Once inside the cell, the esterases in the cytoplasm hydrolyze the esterified fluorophore and the molecule becomes cell impermeant and fluoresces (Tsien, 1989). The plates are then read by a fluorescent plate reader.

MTT [3-(4,5-dimethylthiazolyl)-2-5-diphenyltetrazolium bromide] assays are based on the reduction of MTT by live cells to an insoluble (in aqueous solutions), dark purple formazan precipitate that can be read by a standard plate reader. This is a frequently used high-throughput assay for cytotoxicity (see Chapter 11 for high-throughput format). [^{3}H]Thymidine incorporated into DNA during nuclear replication (or repair) is precipitated by trichloroacetic acid (TCA) and the incorporated counts measured in a scintillation counter. In any of these assays it is important to initially ascertain that cell number correlates with either counts per minute or fluorescence or formazan formation, as there are a number of factors that can influence the outcome. For assays to be performed in microtiter plates, Drummond Scientific manufactures a multichannel aspirator head with a luerlock fitting that can be used with a syringe or vacuum hose. Multichannel pipettors with 8 or 12 tips/pipettor also speed up the task of setting up these assays.

[^{3}H]THYMIDINE INCORPORATION ASSAY FOR DNA SYNTHESIS

Many investigators use [^{3}H]thymidine incorporation as a simple and often rapid determination of cell proliferation. Strictly speaking, this is not a direct measure of cell division, as DNA synthesis is not always indicative of mitosis (Orly and Sato, 1979). It is important not to label the cells for longer than 24 hr so that one is not measuring DNA repair due to strand breaks within the nucleus itself. Care should be taken so that radioactive materials are confined to the immediate work space and disposed of properly. Use disposable plastic beakers for discarded pipette tips and other small items that come into contact with radioisotopes.

Materials

1. 6-well tissue culture plates (or whatever size is most convenient—alter volumes and cell numbers accordingly)
2. [^{3}H]Thymidine
3. Assay medium should be thymidine-free growth medium (2 ml/well). Many nutrient mixtures do not contain thymidine. Those that do can frequently be ordered without thymidine. If the medium is to contain serum, the serum should be dialyzed.
4. Cold 20% TCA. Order a 100-g bottle (Sigma T-4885), carefully add purified water directly to the bottle, and allow to dissolve. Pour the contents of the bottle carefully into a graduated cylinder. Wash the bottle with purified water and add this to the graduated cylinder. QS to 500 ml. Store at 4°C.
5. 0.1 N NaOH
6. 40% acetic acid
7. Scintillation fluid (preferably biodegradable, e.g., BetaMax)
8. Absorbent plastic-backed paper (e.g., Benchkote)
9. Radioactive waste bags
10. 7-ml polypropylene scintillation vials

Procedure

1. Trypsinize the cultures as previously described and take a cell count.
2. Plate 5×10^3 to 1×10^4 cells/well in growth medium, 2 ml/well.
3. Incubate the cultures for 24–72 hr (this will depend on the cell type).
4. Place a sheet of absorbent plastic-backed paper (e.g., Benchcote) in the tissue culture hood.
5. Prepare assay medium: Growth medium containing 1 µCi/ml [^{3}H]thymidine. Dispose of all pipettes and pipette tips in a radioactive waste bag.
6. Remove the plate from the incubator. Remove growth medium from the plate.

7. Add 2 ml of assay medium.
8. Incubate 3–4 hr (this is for an established cell line that has a nominal doubling time of 24–36 hr. Primary cell cultures or cell lines with >36 hr doubling times may require a longer—up to 24 hr—incubation).
9. Remove the radioactive medium. Wash the plate 2× with serum free medium.
10. Add 1 ml/well cold 5% TCA, to extract residual thymidine. *Note:* TCA is a strong acid. Care should be taken when handling the crystals or solutions to avoid getting it on clothes or skin.
11. Repeat step 9. Tip the plate to remove as much TCA as possible.
12. Add 1 ml/well 0.1N NaOH.
13. Leave at room temperature until the cells are unattached and broken up (around 20–30 min).
14. Resuspend the cells in the well with a 1 ml Pipetman. *Note:* The acid–base ratios in the following steps are critical, as well as the order of addition.
15. Add 0.5ml of the cell suspension to a 7-ml scintillation vial.
16. Add 3.5 ml of scintillation fluid.
17. Add 200 μl of 40% acetic acid.
18. Cap the vials and shake.
19. Count in an appropriate beta counter.
20. Dispose of the vials in a bag for radioactive waste, separate from all other radioactive waste, solid or liquid. (Individual institutions usually set up their own guidelines regarding the disposal of biohazard waste and radioactive compounds in compliance with city, state, and federal regulations.)
21. Be sure and treat any spills with solutions such as Count-Off.

HIGH-THROUGHPUT ASSAYS FOR SECONDARY ENDPOINTS FOR CELL NUMBER

It is difficult to develop a high-throughput assay for actually counting cell number. Therefore, many investigators use a secondary endpoint, which usually correlates well with increased or decreased cell number, as a screen. The thymidine incorporation assay described above is one such secondary endpoint. Others are based on colorimetric readouts such as crystal violet, a protein stain, or fluorometric readouts of dyes, such as calcein-AM, whose level in cells is proportional to esterase activity in the cells. Many of these assays are extremely sensitive and therefore can be used with cells in 96-well plates. The results can be automatically read on a plate reader, analyzed by an attached computer, and printed out in a form suitable for publication.

MEASURING CELL VIABILITY

Total cell number both in a tissue *in vivo* and in a culture dish *in vitro* is a balance between the rates of cell growth, or mitosis, and cell death. The ability to differentiate between live and dead cells is therefore important. Trypan blue dye exclusion (described in the section above on using the hemocytometer) is a rapid and reliable method that relies on the breakdown of membrane integrity, allowing the uptake of a dye that is normally membrane impermeant.

The following method is more sensitive in detecting damaged cells than trypan blue and it is quite easy to see the cells and do differential counts (Parks *et al.,* 1979). A fluorescent microscope with an appropriate filter, however, is required. Both acridine orange and ethidium bromide are DNA intercalating dyes and therefore mutagens. Handle and dispose with care.

ACRIDINE ORANGE–ETHIDIUM BROMIDE VIABILITY DETERMINATION

Materials

1. Trypsinized and neutralized cell suspension
2. Ethidium bromide–acridine orange (EB-AC) stock solution (100×): ethidium bromide, 50 mg; acridine orange, 15 mg. Dissolve in 1 ml 95% ethanol and QS to a total volume of 45 ml in purified water. Store in 1-ml aliquots at −20°C.
3. PBS

Procedure

1. Make a working solution of EB-AC by diluting a 1-ml aliquot of stock solution into 100 ml PBS. This can be stored, light tight, at 4°C for up to 1 month.
2. Mix equal volumes of cell suspension: EB-AC.
3. With a Pasteur pipette or a Pipetman pipettor, add a drop (100 µl) to a slide and cover with a coverslip.
4. Observe with an epifluorescent microscope using a filter excitation of 495 nm.
5. Count the live/dead cells with a tally counter. The live cells will fluoresce green and the dead cells will fluoresce orange.
6. Determine viability by calculating: (Live cells/Total cells counted) × 100.

More elaborate and accurate methods of differentiating live and dead cells include FACS, tunnel labeling, and electron microscopy. Methods for measuring apoptotic cell death are evolving very rapidly at this time (Mather and Moore, 1997; Moore *et al.,* 1995, 1996), and more technological advances can be expected.

—————————— PLATING EFFICIENCY ——————————

Plating efficiency will determine how well single cells can survive and form colonies. Since the ratio of the volume of medium to the volume of the cells is very large, there is a minimal impact of the cells on their environment. Thus, there is little opportunity for the cells to metabolize and convert amino acids, to bind toxic components, or to secrete autocrine growth factors they may need for growth. Plating efficiency is thus a more sensitive measure of the cell's response to its surroundings. This may be the method of choice for determining the nutritional requirements of cells or for testing and comparing serum lots or for toxicity testing of compounds. There are many people who prefer this method for assaying the effects of growth factors. However, the optimal concentrations of nutrients and growth factors derived using plating efficiency may be inadequate to support cell growth to high densities. And for some cells the best-known growth conditions are still inadequate to support their survival at low cell density. This indicates that the medium lacks an essential component(s) that the cells are providing for themselves in an autocrine fashion. Fig-

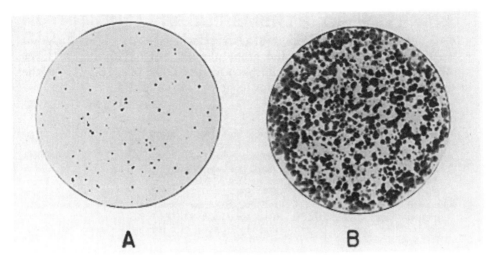

Figure 5.4. The effect of conditioned medium on the plating efficiency of M2R melanoma cells. The two plates were plated at the same density in hormone-supplemented, serum-free medium. The plate at right had 50% of the same medium conditioned for 24 hr by cells at high density. One active factor in the conditioned medium is TGF-α made by the melanoma cells.

ure 5.4 shows the plating efficiency of melanoma cells with and without conditioned medium from high-density cultures of the same cell type.

Plating efficiency is expressed as a percentage:

$$\frac{\text{colonies formed}}{\text{No. of cells plated}} \times 100 = \text{Plating efficiency}$$

Materials

1. Cell cultures
2. Six 100-mm tissue culture plates
3. 15-ml conical centrifuge tubes
4. 60 ml growth medium
5. Trypsin solution
6. Soybean trypsin inhibitor (if serum free)
7. Hemacytometer or electronic particle counter

Reagents for Fixing and Staining

1. 10% Formalin
2. Crystal violet (0.1% in water)

Procedure

1. Trypsinize as described previously. Check visually to make sure the cells are in a single-cell suspension.
2. Neutralize by resuspending in 10 ml of growth medium with serum in a conical 15 ml tube, or add 1 ml STI and resuspend in 10 ml of serum-free growth medium.
3. Wash by centrifugation at 900 rpm for 3–4 min. Resuspend in 5–10 ml of growth medium.

4. Count an aliquot of cell suspension.

5. Determine the volume needed to aliquot 100, 500, and 1000 cells to duplicate 100-mm plates. Because these volumes can be necessarily small, it is convenient to dilute to the appropriate cell number in 1 ml, and add this volume to 9 ml of growth medium.

6. Resuspend the cells thoroughly in the 10 ml of medium by repeated pipetting to make sure the cells are dispersed. Plate the cells in an appropriate growth medium in the 100-mm plate.

7. Incubate the plates until the colonies are large enough to be visible to the naked eye (0.2–1 mm in diameter) but not running into each other. The time required will depend on the cells but is usually around 10 days.

8. Wash the plates 1× with PBS; remove the PBS and add enough 10% formalin to cover the plate. Fix for 10 min.

9. Remove the formalin and add enough crystal violet solution (2–3 ml) to cover the plate. Leave 10 min, remove the crystal violet, and rinse under running water until no color is left.

10. Count the colonies with a marker pen or use a colony counter if available.

CONDITIONING MEDIUM

Conditioned medium can be used to grow the same cells at low density that may ordinarily require a high density to grow (for example, see Fig. 5.4). One also may use conditioned medium from one cell type to grow another type of cell in continuous or primary culture. In theory, one is allowing the conditioning cell culture to provide the growth factors or metabolic products needed by the culture grown in the conditioned medium. Conditioned medium is prepared using fresh medium in the following manner:

1. Allow a plate of the cells to be used to condition the medium to grow to near (at 70–90%) confluency.

2. Wash the plate once with medium. Then, fresh growth medium (± serum or hormone supplementation) is placed on the cells for 24–48 hr. If the cells are very dense or metabolize the medium rapidly, then use the shorter time period for conditioning. The conditioned medium should not be very yellow (i.e., have a pH < 6.9) when removed from the cells. If the pH is too low, it indicates that the medium is exhausted and may therefore either contain a lot of debris and released proteases from dying cells in the original culture and/or not contain sufficient nutrients to support the growth of fresh cells.

3. This medium is collected, filtered through a 0.22-mm low-protein-binding filter to remove any floating cells and used for "conditioned medium" where described.

4. This medium can generally be stored in a polypropylene tube (to minimize binding of factors to the tube) at 6°C for 1–2 weeks.

5. If the cells to be grown in conditioned medium are cultured in serum-free medium, combine the conditioned and fresh medium (mixed at 10–50% v/v) and add the necessary hormones and growth factors based on this total volume.

CLONING

The purpose of cloning is to assure that all cells in the culture are descended from a single cell, that is, genetically identical. This prevents the rapid and unpredictable changes in culture phenotype that may occur in mixed cell populations when conditions change to favor one cell type over another. Recloning established cell lines will frequently give a line with more consistent characteristics. If one wishes to change the cell's properties through mutation or transfection, cloning is necessary to assure that all cells in the study population are similar. However, all cell cultures, cloned or not, are subject to genetic drift. Therefore, there may be significant variation in a specific phenotype after only a few passages if selective pressure is not maintained.

Additionally, cloning allows the screening of a large number of cell lineages and the selection of a cell line with the desired characteristics. Since most transfection and mutagenic techniques yield cells with widely differing genetic changes and growth rates, cloning allows one to choose a cell strain with the optimal properties for study. There can be considerable change with time, however, even in cloned populations. These changes can be genetic, and therefore irreversible, or a phenotypic response to changing or marginal culture conditions that can be controlled or reversed.

Cloning for a homogeneous cell type can be achieved in a manner similar to setting up cells for plating efficiency. Usually, 100–1000 cells can be seeded and dispersed well enough in a 100-mm plate so that many widely separated colonies may be observed and selected after about 10 days in culture. Twenty to 100 colonies per 100-mm plate generally gives a good separation so choose the initial cell number plated based on the plating efficiency of the cell line being used. Several plates can be seeded at different densities and the best plate picked for cloning. Do not clone from cultures where the cells have an abnormally low plating efficiency (e.g., <10%), since these might not be representative of the starting population. For cultures in early passage or for cell lines difficult to grow at low density, the addition of 10 or 20% conditioned medium may promote colony formation. If the cells to be cloned normally grow in suspension or move around a lot on the plate (that is, tend to grow in very dispersed colonies with a lot of single cells between rather than compact ones), use the cloning-by-limiting-dilution method described below.

CLONING BY PICKING COLONIES OF ATTACHED CELLS

Materials

1. Cell cultures
2. 100-mm plates
3. Growth medium and conditioned medium, if necessary (see below)
4. Hemacytometer or electronic particle counter
5. Stainless steel cloning rings (sterile)
6. Silicon grease (sterile). Do not use stopcock grease. Place the grease in a glass petri dish and spread a layer about 2 mm thick over the surface. Place the cloning rings in the grease and autoclave (Fig. 5.5).
7. Microscope ring marker
8. 100-μl pipettor and sterile tips

Procedure

1. Trypsinize cultures as previously described.
2. Resuspend in 5 ml of growth medium.
3. Count an aliquot of cells.
4. Prepare a suspension to yield 500–1000 cells/ml.

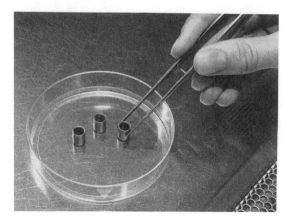

Figure 5.5. Stainless steel cell cloning rings can be autoclaved and reused. The weight of the steel rings makes them easier to handle and less likely to move or leak than plastic cloning rings.

5. Add 1 ml of cell suspension to each of two to four 100-mm plates; pipette cells repeatedly to disperse.
6. Incubate and observe starting on day 5.
7. When colonies contain 50–500 cells, they can be picked; make sure that the colonies selected are far enough away from all their neighbors so that the cloning ring will encompass only one colony.
8. Aspirate the medium from the plate and wash with 5 ml of PBS.
9. Remove the cloning ring from the silicon grease with a forceps and place it over the colony; press it gently with the forceps to assure a good seal. The cloning ring should not contain more than one colony. Check this visually.
10. Continue until you have 5–10 colonies/plate selected and covered.
11. With a 100-μl pipettor, add 25 μl trypsin solution to each cloning ring. Incubate until the cells are rounded up and coming off the plate when checked under the microscope.
12. Using a 100-μl pipettor, dilute the trypsin with 75 μl of growth medium and carefully pipette up and down, transferring the suspension to one well of a 24-well plate.
13. Add 400 μl of growth medium to the well.
14. Fill any empty wells with sterile water or PBS.
15. Incubate at 37°C and monitor growth daily.

CLONING IN SERUM-FREE MEDIA

Cloning in serum-free medium is similar to the previous cloning method except that it is necessary to neutralize the trypsin with soybean trypsin inhibitor and then wash by centrifugation. Because this requires more handling of the cells, let the colonies grow somewhat larger than when cloning in serum-containing medium. It is more likely that conditioned medium will be required for low-density growth. In this case, condition the serum-free medium, then add fresh growth factors when the conditioned medium is used.

Materials

Same as above

Procedure

1. Trypsinize as previously.
2. When cells have rounded up, add 25 μl of a 1 mg/ml solution of soybean trypsin inhibitor.
3. Carefully pipette the suspension with a 100μl pipettor; transfer to a 15-ml conical centrifuge tube.
4. Add 5 ml of medium, centrifuge at 900 rpm for 3–4 min.
5. Carefully aspirate the supernatant as the pellet will not be visible.
6. Using a 1-ml pipettor, resuspend the pellet in 500 μl of growth medium and transfer to one well of a 24-well plate containing all necessary supplements.
7. If an attachment factor is required for growth, precoat the wells.
8. If conditioned medium is to be used, make up fresh medium with twice the final desired concentration of supplemental factors and mix 50/50% with conditioned medium.
9. Add sterile water or PBS to any unused wells.
10. Incubate and monitor daily.

CLONING BY LIMITING DILUTION

Alternatively, cells may be cloned by the limiting dilution method. This method can be used with suspension or attached cells. It is the only method that should be used with suspension cells or with cells that are very mobile when attached. Because the cells are plated at a low cell–medium volume ratio, conditioning of the medium (see above) is frequently necessary to get good growth.

Materials

1. Cell culture
2. 96-well microtiter plate
3. Growth medium (and conditioned medium)
4. Hemacytometer or electronic particle counter
5. Trypsin–EDTA solution

Procedure

1. Treat the plate of cells to be cloned with trypsin to obtain a single-cell suspension. Visually inspect to ensure that most cells in the suspension are separate.
2. Wash cells twice with medium by centrifuging at low speed.
3. Take cells up in a complete growth medium to a final density of 10–100 cells/ml.
4. Plate at 100 μl/well in a 96-well plate. The density used will depend on the plating efficiency of the cell line used. To maximize the chances of having a single cell/well, approximately half of the wells should be empty.
5. Check colonies visually 12–24 hr after plating and mark wells or colonies that arise from a single cell. If two cells are seen in a well, it should be discarded.
6. Clones can be initially screened directly from the 96-well dish using whatever assay is appropriate. For example, suppose we are screening clones for those that produce high levels of a recombinant protein after transfection. If an antibody is available for a Western blot, then the clones can be screened by transferring an aliquot of medium directly onto a nitrocellulose filter in a 96-well manifold. The filter can then be handled as for a

normal Western blot. Cells from the wells that give the strongest reaction can be chosen for further expansion. If one is screening hybridomas for antibody production, this too can be done by collecting medium directly from the 96-well plate, as long as the clones are grown in an antibody-free medium. When more cells are needed for a screen, clones can be picked and grown before screening. One can screen for improved cell growth by direct visual observation. Make sure that all wells are visually checked on day 1 to make sure that there is only one cell/well.

7. When clones are picked, they should be passaged first into a well of a 24-well dish, then to 35-mm, and finally to 100-mm plates. Conditioned medium may also be used in the first passage after cloning, if necessary.

8. To increase the statistical probability of obtaining a single-cell clone, cloning should be repeated two to three times in succession.

The medium used should be the normal maintenance medium (e.g., F12/DME plus 7.5% FBS; see Chapter 4, medium selection section). The addition of 5 μg/ml insulin during the low density growth required before cloning and during the first subsequent passage is helpful for many cell types. Conditioned medium can be mixed with normal growth medium (1:1 v/v) if the cells are difficult to grow at low density. If the cells are being cloned after transfection and selection, remove the selective agent (e.g., G418, methotrexate) from the medium during cloning. After the cells grow to higher density, the selective agent can be returned to the medium. This is less stressful to the cells and allows a higher plating efficiency and better growth during the cloning steps.

——— FREEZING AND THAWING CELLS ———

The ability to freeze and preserve cells in liquid nitrogen for many years with minimal loss of viability is one of the advantages of working *in vitro*. Having a frozen bank of cells provides a backup in case cells are lost due to contamination, carelessness, equipment failure, or a natural disaster. Some types of primary cultures can be prepared in large batches, a large number of vials frozen, and the cells thawed sequentially and studied as secondary cultures so that a large number of experiments can be performed on early passage cells from the same preparation. Alternatively, as one tries to establish a cell line (see Chapter 10), a few vials of the cells should be frozen every three to five passages to have a permanent record of any changes that may occur with passage number. For normal human cells, which have a limited life span *in vitro*, expansion and freezing of an early passage bank is the only method that allows similar cells to be used in many different laboratories. Whenever a new property of a cell line is thoroughly characterized or a cell line recloned, a number of vials should be frozen at that point in order to be able to return to these cells if the lines being carried change.

In industrial production using mammalian cell lines, very large cell banks are prepared and thoroughly characterized as to the absence of adventitious agents and the cell properties (see Chapter 12, cell banking section). Cells thawed from these banked vials are used for a strictly limited and carefully defined period to avoid any alteration in the cells during culture.

Cells are frequently easier to ship as frozen vials rather than in flasks. The American Type Culture Collection (ATCC), a repository of several thousand cell lines, would not be possible without the ability to bank and store cells frozen.

However, there are some cautions that should be observed when freezing cells. If the majority of cells die during the thawing process, then the resultant cell population may not be the same as that frozen. This may be a major issue if one is freezing and thawing primary cultures containing mixed cell types. The process of freezing and thawing may alter cell karyotype, particularly those of rat or mouse cells. Thus, freezing may generate heterogeneity in a cell line. If cells are stored in the liquid phase of liquid nitrogen, this can be a source of vial-to-vial contamination, either cross-cell contamination or spread of adventitious agents such as mycoplasma. Finally, it should be recognized that cells stored at $-180°C$ will still, albeit very slowly, lose viability. Therefore, valuable lines should be periodically thawed, expanded, and refrozen to insure a continued supply of cells.

FREEZING

Cells are best frozen as a cell suspension. Always use a healthy culture to provide the stock to freeze cells. A good standard is to freeze $1–3 \times 10^6$ cells/2 ml vial. This is generally sufficient to start a 100-mM plate on thaw and have sufficient cells to work with within two to three passages of thaw. If very large volumes of cells will be needed or the cells are to be used directly from thaw, cells may be frozen at $1–5 \times 10^7$ cells/5 ml freezing vial. Since the cells do continue to metabolize medium components at a much reduced rate while frozen, they should not be frozen at too high a density.

Materials

1. Freezing medium. This should be 90% normal growth medium and 10% glycerol or 10% dimethyl sulfoxide as a cryoprotective agent
2. Freezing vials, 2 ml or 5 ml
3. Freezing container

Procedure

1. Remove the cells from the plate using trypsin or whatever agent is generally used to subculture the cells.
2. Wash the cells once in fresh medium by gentle centrifugation.
3. Resuspend in the freezing medium at the desired cell density.
4. Aseptically pipette the cell suspension into freezing vials. Freezing vials can be polypropylene, with a flexible gasket or glass. Glass vials must be flame sealed and then leak tested by immersion in a dye bath before freezing. Leaky vials not only might lead to cross-contaminated cultures, they might explode on thawing, as the nitrogen that entered during storage rapidly becomes a gas and expands. While cells frozen in glass vials will not leak unexpectedly, and therefore cross-contaminate the culture, this is a more laborious and time-consuming method; most small laboratories now use polypropylene freezing vials. *DO NOT* freeze or store cells in liquid nitrogen in any kind of container other than those designed for frozen cell storage. Many plastics will become brittle at liquid nitrogen temperatures and shatter or explode when removed, possibly injuring laboratory personnel.
5. Label *EACH* vial individually with the designation and passage number of the cells, the date of freezing, and the initials of the person freezing the cells. It is important that each vial be labeled, not just one label placed on the box or cane. Boxes and vials can be dropped, and more often than not, only those individually labeled will be salvageable. Many freezing vials have an area of the surface suitable for writing on with an indelible (make sure it is indelible in organic solvents and liquid nitrogen) ink pen. If labels are

to be taped or glued on, make sure the adhesive will withstand liquid nitrogen temperatures. There is nothing quite so frustrating as a freezer full of vials of your precious collection of frozen cells, all of whose labels are at the bottom of the tank.

6. Cells should be frozen slowly. A steady decrease of one degree per minute is ideal. Since that means a temperature change of 220°C, it should take a minimum of 4 hr to reach liquid nitrogen temperatures. There are machines that will freeze vials at this exact rate. However, this is clearly not a required expense for the average cell culture laboratory. The ideal rate can be approximated adequately using one of the following techniques:

   • Vials of cells should be placed in a freezing rack or a polystyrene plastic (Styrofoam) tube rack that fits snugly. The best commercially available freezing containers have a space that can be filled with isopropanol (see Fig. 5.6). The rack or container should be placed in the refrigerator for 30 min, then moved to a −80°C freezer, where it should remain for at least 60 min and no longer than overnight. The vials may be transferred from the freezing container to the liquid nitrogen freezer.

   • Alternatively, nitrogen tank manufacturers make inexpensive freezing necks that fit their standard tanks. This replaces the insulating cap and holds the vials high in the neck of the tank, allowing them to be frozen slowly by the nitrogen vapors.

Cells may be stored for a short time at −80C, but they will continue to degenerate rapidly at this temperature and so should not be stored at −80°C for more than a few weeks. A selection of storage systems is shown in Fig. 5.7.

TEMPORARY FREEZING OF LARGE NUMBERS OF CLONES

If large numbers of cell clones, for example, hybridomas or transfectants, are being screened, there is a fast way to temporarily freeze large numbers of colonies. Set up the clones in two identical 96-well plates (one plate can be split into two by using a multiwell pipettor), place freezing medium in one, seal the edge of this plate with tape, place it in a Styrofoam container so that it will cool slowly, and then after 30 min in the refrigerator place it in the −80°C freezer. It may be stored in this fashion for 2–3 weeks while the other plate is screened. There is then a backup plate available in case of an unfortunate accident, or if one later wishes to rescreen a clone that was discarded. The entire plate can be thawed, the freezing medium replaced with normal medium, and the plate returned to the incubator. Cells stored in this fashion will remain viable only for a few weeks, so they should be discarded after this time.

Figure 5.6. Equipment for freezing cells slowly so that they reach liquid nitrogen temperature at the optimal rate.

Figure 5.7. Equipment for freezing and storing cells at liquid nitrogen temperatures. There are three different-sized tanks in the photo. The large square tank in the right rear has an automatic fill feature. As long as the liquid nitrogen tank on the left is replaced regularly, this tank will not go dry. The tank on the left requires manual filling but has a long holding time (long time between required fills). Both tanks have a box inventory system. The small tank on the right front uses tubes taped onto canes for storage, making retrieval of a given vial more difficult if many different cell types are stored in the same tank.

THAWING

Cells should be thawed as rapidly as possible, in contrast to the optimal slow rate of freezing. In both instances, this is to minimize ice crystal formation, which may damage cells.

Procedure

1. A vial of cells should be removed from the liquid nitrogen tank using insulated gloves and handling tongs.
2. The vial label should be carefully checked and the cap checked to insure that it is still tightly screwed shut.
3. The entire vial can be immersed in a 37°C water bath and gently shaken until thawed. *Note:* Be aware that freezing vials can explode on thawing. This is one reason many people use plastic rather than glass vials. However, take care with both.
4. The contents of the vial should be transferred to a sterile 15-ml tube containing warmed medium and the cells washed by pelleting with gentle centrifugation (800 rpm, 3–5 min).
5. The freezing medium is then discarded and the thawed cells resuspended in fresh culture medium.
6. Plate the cells in their normal growth medium. The initial plating density should be high (at $2–3 \times 10^4/cm^2$). The cells should be placed in the incubator. When this culture is

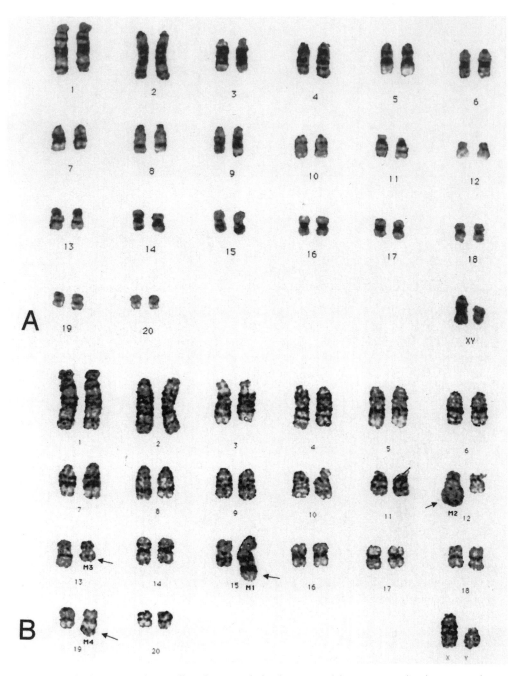

Figure 5.8. The karyotype of ESC cells unfrozen and after freezing and thawing. Note the chromosomal rearrangements (marked by the arrows) that occurred on freezing and thawing. (A) Cells carried continuously for 40 passages in serum-free, hormone-supplemented medium. (B) Cells frozen at passage 10 then thawed and carried to passage 20 in the same medium.

confluent (1–3 days), then the cells should be split at a high ratio (1:2–1:10) for the first passage after thaw. After the first passage, most cells are fully recovered from freezing and can be handled normally.

As with cloning, it is important to have a good recovery of cells from the frozen vial (preferably > 90%). If most of the cells are lost on thawing, there is a real possibility that the characteristics of the thawed culture will differ from those of the line frozen. Allow at least a few days (and preferably two passages) postthaw before using the cells in an experimental protocol. It should be noted that freezing and thawing can change the karyotype of cells, as shown in Fig. 5.8.

FROZEN CELL STORAGE

Frozen cells should be stored at −180°C. This is liquid nitrogen temperature. Cell storage tanks, made by several manufacturers, are specially designed to handle these low temperatures. A good tank should have a long "holding time" (the amount of time it takes all the liquid nitrogen to evaporate if the tank is not refilled), hold a reasonably large number of vials, and have a good inventory system for storing and finding the vials. Larger tanks use less liquid nitrogen in proportion to their storage capacity than smaller tanks. Optimally, cells should be stored in the vapor phase rather than the liquid phase, but this increases the risk that the tank will go dry. Some type of alarm or an automatic filling system is very helpful.

If the work of the laboratory depends on making new cell lines or clones that are not available elsewhere, it is wise to store frozen vials of these cell lines in at least two separate freezers so that everything is not lost if one goes dry because of carelessness or a defect in the tank. Remember that this equipment, though usually stuck away in a corner, contains the heart of a cell culture laboratory and may contain irreplaceable cell lines and cultures that are stored nowhere else. Additionally, the vials are placed in the tank and thereafter are out of sight, hard to get to, and difficult to inventory. Frozen vials of cells frequently stay in a laboratory long after the student, fellow, or technician who froze them and placed them in the tank is gone. Therefore, a good record-keeping system is an essential part of cell storage.

RECORD KEEPING

Since by its nature a frozen vial of cells is something that will be put away and not looked at for a long time, perhaps years, labeling and record keeping are critical to maintaining a good laboratory cell bank. Each vial of cells should be labeled with the complete name of the cells, the passage number, the date, and the name or initials of the person growing and freezing the cells. A laboratory-wide system should be established to maintain records of what cell lines are frozen, how many vials are frozen, who froze them, why they were frozen at that time, and where they are stored. Records should also be kept when vials are thawed so that it is clear when the banked vials for a given line are running low. These may be done on the same form. A simple solution is to draw up a form such as that shown in Fig. 5.9 and keep these in a notebook near the cell freezer. Laboratory personnel can fill out a page every time they freeze cells, and vials can be marked off when they are removed from the freezer. The notebook can be divided into sections for the different cell types in the freezer. Alternatively, if there is a laboratory computer conveniently available, the records can be kept in a database on the computer. Any system is only useful if it is used by

Freezer ID:		1	2	3	4	5
Rack:						
Box						
Cell Line		6	7	8	9	10
Cell Type						
Date Media:						
Frozen in		11	12	13	14	15
Growth						
Density:		16	17	18	19	20
#Vials						
Location		21	22	23	24	25
Ntbk ref.:						
Total:						

Notes:

Figure 5.9. Copy of forms used for tracking frozen cells in a liquid nitrogen freezer. The form on the left records relevant information about the cells frozen and the right-hand form gives location.

all members of the laboratory. Alternatively, one laboratory member can be responsible for placing cells in the freezer and removing cells to be thawed. This person would then keep the freezer records.

SUMMARY

This chapter has outlined the basic techniques that anyone who wishes to do cell culture at any level will need to master. Knowing how to subculture, clone, freeze and thaw cells, and how to do growth curves and plating efficiency experiments will allow the student or scientist to ask sophisticated questions about growth regulation of cells *in vitro*. These techniques, coupled with an understanding of the function of the media components and the control of the physical environment, will provide a good basis for setting up and overseeing a tissue culture core facility and for troubleshooting problems.

This information is also critical for those who wish to use cell culture as a tool to study growth factor signal transduction, clone or express genes, or ask questions concerning the regulation of development *in vitro*. If the cells used in these studies are not healthy and the culture techniques reliable and reproducible, then the investigator runs the very real risk of studying artifacts. Unlike many molecular biology techniques where it is possible to tell at the end of a series of manipulations if there has been a mistake (e.g., "Do I have the right gene?"), with cell culture this retrospective checking is usually impossible. For example, by the time the cells have been harvested and the mRNA extracted, it is too late to ever know whether that culture was in log or lag phase, was contaminated with other cells or mycoplasma, or was phenotypically normal. This information must be obtained by monitoring the cultures to be used continuously during routine passage and during experiments.

The following two chapters on looking at cells and dealing with contamination round out the basics of cell culture. The final five chapters deal with more specialized areas of cell culture, from serum-free culture to an introduction to the use of cell culture in commercial settings.

REFERENCES

Mather, J., and Moore, A., 1998, Culture media: Large-scale production of proteins in animal cells, in: *The Encyclopedia of Bioprocess Technology: Fermentation, Biocatalysis, and Bioseparation* (M. Flickinger and S. Drew, eds.), John Wiley, New York, (in press).

Moore, A., Donahue, C. J., Hooley, J., Stocks, D. L., Bauer, K. D., and Mather, J. P., 1995, Apoptosis in CHO cell batch cultures: Examination by flow cytometry, *Cytotechnology* **17**:1–11.

Moore, A., Mercer, J., Dutina, G., Donahue, C., Bauer, K., Mather, J. P., Etcheverry, T., and Rhyll, T., 1997, Effects of temperature shift on cell cycle, apoptosis and nucleotide pools in CHO cell batch cultures, *Cytotechnology* **23**:47–54.

Orly, J., and Sato, G., 1979, Fibronectin mediates cytokinesis and growth of rat follicular cells in serum-free medium, *Cell* **17**:295–305.

Parks, D. R., Bryan, V. M., Oi, V. T., and Herzenberg, L. A., 1979, Antigen-specific identification and cloning of hybridomas with a fluorescence-activated cell sorter, *Proc Natl Acad Sci USA* **76**:1962–1966.

Tsien, R., 1989, Fluorescent probes of cell signalling, *Annu. Rev. Neurosci.* **12**:227–253.

CHAPTER 6

Looking at Cells

Some of the most powerful methods of gathering data in the cell culture laboratory use visual observation. While it can take a good deal of time to develop techniques to transfer the visual observations into quantitative data, these numbers frequently still do not convey the qualitative information conveyed by visual observation. Visual observation thus remains an invaluable tool of the experienced cell culturist and cell biologist in gathering information on the status of cultures and experiments. This information is extremely valuable in making the ongoing decisions involved in establishing cell lines, maintaining healthy cultures in the laboratory, and deciding which lines of research to pursue. A trained eye and the knowledge base to interpret the observed phenomenon are therefore assets to any cell biologist.

JUST LOOK AT THE DISH

Some information can be obtained from simple visual observation of cultures. What is the pH (the color, if the medium contains a pH indicator)? Is it neutral, acid, or basic? If the pH is above 7.5 (purplish color), there is a problem with the incubator. Check to make sure that the CO_2 supply is not empty. Check the CO_2 setting to make sure it is correct for the medium being used. If both are correct, check the CO_2 level in the incubator directly to see that it is, in fact, what the readout says it is. If the medium is very yellow, the medium is acid. This can indicate a very dense cell culture, too long a period between subculturing or feeding the cells, or the growth of microbiological contaminants. Look at the clarity of the medium. Is it cloudy? If so, check carefully for contamination or debris from dead cells. Look at the cells on the plate. Is the monolayer lifting off or unevenly distributed? All this information, which can be obtained in a few seconds of observation a day, is important in monitoring the health of the cultures.

THE LIGHT MICROSCOPE LEVEL

The most important tool for visual observation of cultures is the light microscope. All cultures should be visually observed at 50 to 500-fold magnification before the quantitative analysis of cell counts, mRNA measurement, or any other experimental procedure. Any change in cell morphology should be noted in the notebook in detail. A general impression of cell number relative to control can also be noted. Sometimes, one may observe a marked change in cell size, shape, or cell–cell association that will not be reflected in a change in cell number (or a change in some other specific parameter being quantified). These can provide important leads for future experiments. The best way to record these changes is photographically. An instant camera (e.g., Polaroid) is useful for taking the occasional photograph to record a morphological change. For more extensive photo sessions, 35-mm film is a less expensive alternative and will provide better quality publication prints. Alternatively, 35-mm slides can be stored and viewed easily and can be used directly in presentations to large groups. However one chooses to take the photograph, it is essential to have a good system for maintaining a record of what the photographs represent. One cell might look very much like another to the print shop, or even to the photographer a year later. If no camera is available, a quick sketch in the notebook can often convey relevant features. Several different types of microscopes and their uses are described below. Inverted microscopes are preferred for viewing living cultures, since having the objectives under the stage and the condenser on top permits viewing cells in culture dishes or flasks (see Fig. 6.1).

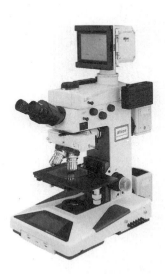
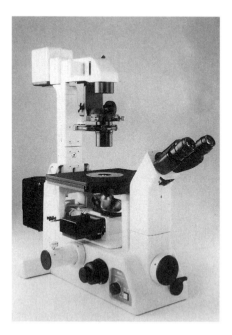

Figure 6.1. Photographs of upright and inverted microscopes. The upright microscope on the left is equipped for bright field, phase contrast, and fluorescence microscopy. The exposure meter is built into the microscope base. The microscope at right is an inverted phase contrast microscope with a camera back-attached to the front port.

PHASE CONTRAST

Phase contrast is a system invented in 1932, by F. Zernike, for which he won the Nobel prize in physics. The phase differences occur by placing a phase annulus in the objective and mounting a fixed annular phase ring in the condenser system. This configuration creates a ring of light in the rear focal plane of the objective, where 75% of the central beam is absorbed. The refracted beam of light (created by varying thickness and refractive index of the features of the cell) is brought into focus at the eyepiece, where it produces a bright image. Because the intensity of the beam is significantly reduced in the rear focal plane of the objective, the result is a contrasting dark background for the bright image, described by its "phase contrast." The inverted phase contrast microscope gives much information when viewing unstained living tissue (Fig. 6.2). In a typical flattened cell, the nucleus, nucleolus, large vacuoles and vesicles, lipid droplets, and the cell processes can be easily seen, revealing details in specimens possessing only slight differences, in refractive index, from the surrounding medium. Several phase contrast photos of different kinds of cells are shown in Figs. 6.3 through 6.5.

The first microscope then to be bought for general use in the cell culture laboratory should be an inverted microscope fitted with phase contrast optics. It should be chosen with future upgrades in mind, such as epifluorescent illumination, Nomarski (differential interference) or (modulation) contrast illumination, or dark-field illumination, which requires only the addition of the necessary optical components and obviates the need for purchasing a second or third microscope. It should offer several ports for the attachment of video and

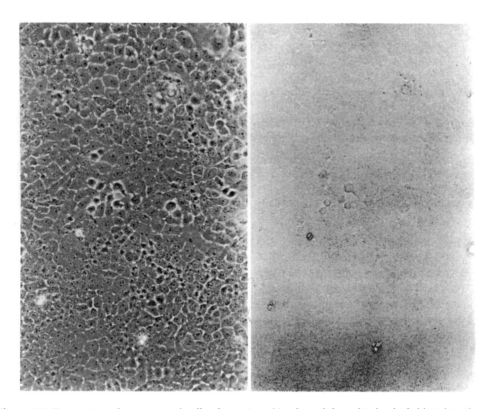

Figure 6.2. Comparison of an unstained cell culture viewed in phase (left) and in bright field (right). The phase contrast optics allow much more detail to be seen.

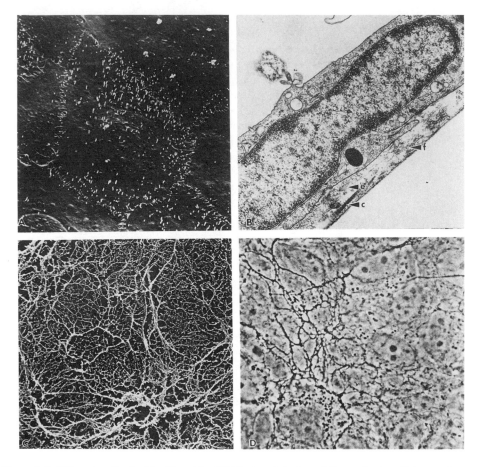

Figure 6.3. Different methods of looking at the TR-M rat peritubular cell line and extracellular matrix made by the cells. (A) The top of a confluent cell monolayer (SEM). (B) A transmission electron micrograph of the same cells sectioned perpendicular to the monolayer. Extracellular matrix material (arrows e,f), including collagen fibrils (arrow c), can easily be seen beneath the cells. (C) A scanning electron micrograph of the matrix after removal of the cells by ammonium hydroxide treatment. (D) A high-power phase contrast photo of the monolayer in which the plane of focus is on the matrix material beneath the cells. Cell nuclei can be seen out of focus.

still photographic equipment, as well as devices for photometry and image analysis. For the teaching laboratory, inexpensive inverted phase contrast microscopes such as the Nikon TMS (see Fig. 2.5) can be used. The visual impact of viewing living cells in phase is so important that every effort should be made to purchase at least one such microscope for a teaching laboratory.

Objectives

There are several factors to consider when choosing objectives. Perhaps the first consideration is resolving power, or that property that allows two components or structures in close proximity to each other to be seen as distinct and separate objects. This property is

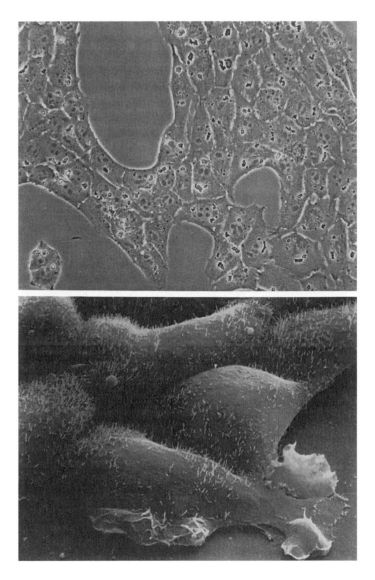

Figure 6.4. (A) Phase contrast and (B) scanning electron micrograph of the TM3 mouse Leydig cell line.

directly related to the numerical aperture (NA) of the objective. The higher the NA, the higher the resolving power, and thus the greater detail can be observed. Another factor to consider is whether the objective will be used for phase contrast only, or for fluorescence only, or a combination of both. Third, consider whether a long-working distance (LWD) condenser is required. For observing cultures grown in Corning 25-cm² flasks, for example, a minimum working distance from stage to condenser assembly is 30 mm. For larger flasks (Corning 175 cm²) a minimum distance of 40 mm is needed. The disadvantage in using LWD condensers is their NA. The NA of the condenser should nearly match the NA of the objective for maximum resolution and image quality, and higher NAs can only be

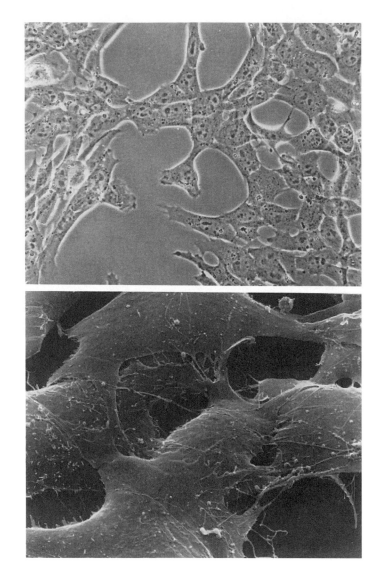

Figure 6.5. (A) Phase contrast and (B) scanning electron micrograph of the TM-4 mouse Sertoli cell line. Note that while the cells in this figure and Fig. 6.4 look similar under phase, they have very distinct morphologies in scanning electron microscopy.

achieved with shorter working distance condensers (< 20 mm). Nomarski differential–interference contrast (DIC) microscopy also requires fairly short working distances, where the NA of the DIC prisms must exactly match the NA of the objective. The numerical aperture is engraved on the barrel of the objective. In addition, the objective will have the eyepiece tube length and coverslip thickness engraved as well. Many objectives will indicate a coverslip thickness of 0.17 mm. This would translate into a number 1-1/2 coverslip; when used on a slide viewed on an inverted microscope, the coverslip should be facing the objective. If the objective has 1.2 engraved on the barrel, the slide should be viewed on an inverted microscope with the coverslip facing up, as 1.2 mm is the thickness of the standard glass slide.

HOFFMAN OR NOMARSKI OPTICS

For all practical purposes, the images produced by these two systems are equivalent. They both convert phase gradients into intensity variations (a form of optical shadowing) so that the cell and its subcellular components can be observed as three-dimensional images in vivid detail (Fig. 6.6). We use the Hoffman modulation contrast system in our laboratory for the following reasons: (1) We can use plastic tissue culture dishes. Plastic is a birefringent (double refractory) material. The polarization techniques required by the Nomarski system preclude the use of plastic between the objective and condenser, as this will significantly degrade the image. (2) An LWD condenser can be used; the Nomarski systems have a shorter depth of field. (3) The image contrast can be varied to control the scattering effect of light as it passes through differently sized and shaped cells and structures in the monolayer. The setup time involved is far less with the Hoffman modulation contrast system so that if a dedicated microscope is not available, an existing microscope can fairly easily be converted, and normal phase viewing is still possible. The purist may still opt for Nomarski optics, as this system can produce extremely high-resolution images, so it is recommended that the investigator try to work with both systems to evaluate the practical potential of both before deciding on one or the other.

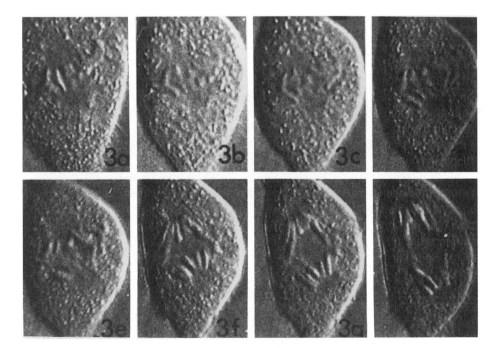

Figure 6.6. Nomarski images of a cell undergoing division. The condensed chromosomes and other cell organelles can easily be seen using the Nomarski optics in these photographs, since this cell remains rather flattened during division.

CARE AND HANDLING OF THE PHASE
CONTRAST MICROSCOPE

Adjusting the phase contrast is a rather simple technique that often is assiduously avoided for no good reason. Conversely, there are the uninitiated among us who will, with the best of intentions, badly misalign the phase rings (phase annular diaphragm) and fail to center the condenser, thus rendering the principal function of the microscope useless (Fig. 6.7). For those readers who have an interest and no manual is readily available, we offer the following steps for keeping the microscope in pristine alignment. This is illustrated using the Nikon inverted microscope, so placement of the adjusting screws will differ depending on the microscope used, but the theory should be similar.

1. Place a dish of cells on the microscope stage and focus using the 10× objective.
2. Narrow down the field aperture diaphragm until the entire diaphragm can be seen in the field of view (Fig. 6.8).
3. Rotate the condenser knob until the aperture edge is in focus.
4. Adjust the condenser centering screws until the field aperture diaphragm image is in the center of the field of view.
5. Open up the field aperture diaphragm.
6. Set the turret assembly (on the eyepiece tube) to "B". If you are using the older Diaphot, remove an eyepiece and replace it with a centering telescope.
7. You will see a phase plate image (dark) and an annular diaphragm image (illuminated) (Fig. 6.8). The idea is to center those two images.
   a. Diaphot 300. This has two centering screws that you adjust by inserting two 1-mm Allen hex wrenches or the Allen hex wrenches that Nikon provides with their microscope (and that sooner or later everyone misplaces) into the hexago-

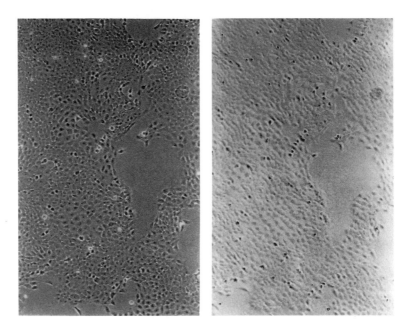

Figure 6.7. (Left) In-phase and (right) out-of-phase image of a cell culture. For a phase contrast microscope to be useful, the phase must be kept well adjusted.

nal screws on the rotating condenser turret. Make sure the condenser diaphragm
is fully open.

b. Diaphot TMD. This condenser has two knurled centering screws located near
the rear of the condenser turret (not to be confused with the two knurled knobs
on the front of the turret for centering the condenser). Adjust these as above to
center the two images. In theory, these things should rarely have to be done (unless there is a compulsive knob-twister in the laboratory). But in this imperfect
world you will find that a practical knowledge of these things can save a lot of
marriages, friendships, and professional collaborations.

FLUORESCENCE MICROSCOPY

As mentioned earlier, an inverted phase contrast microscope can be fitted with epifluorescent (derived from the term *episcopic,* or incident-light) illumination, which will greatly expand the capabilities of the microscope. If a great deal of fluorescence microscopy is
anticipated for visualization of fixed cells, an upright microscope, dedicated to fluorescence, is desirable. The upright epifluorescent microscope can also use oil immersion, high-
magnification objectives if very high magnification is needed for detecting microorganisms
such as mycoplasma. Immunofluorescence has been the most common application of fluorescence microscopy, in that it combines the specificity and spatial resolution of fluorescence microscopy with the selective binding of antibodies to their respective epitopes. By
choosing fluorophores of different colors, multiple sites can be localized on the same cell.
Moreover, fluorescent probes can be incorporated into living cells to localize, visualize, and
measure a wide variety of physiological changes.

The components of an epifluorescent system consist of a light source, usually mercury,
a filter block containing an excitation filter, a barrier filter and a dichroic mirror, and suitable objectives for fluorescence. The excitation filter is a color filter that transmits only
those wavelengths of the illumination light that excites a specific dye. The emission or barrier filter is a color filter that attenuates all the light transmitted by the excitation filter and
transmits any fluorescence emitted by the cell. The dichroic mirror is set at a 45° angle to
the optical path of the microscope. Its coating has the ability to reflect the excitation light
and transmit the fluorescence.

Both organic and nonorganic substances can exhibit some fluorescence, known as
autofluorescence. Autofluorescence is often a major source of unwanted light in the observed fluorescent image and can be minimized in fixed cells by paying careful attention to
the fixation method to be used. For example, glutaraldehyde will fluoresce at wavelengths
well into the UV range, as will a number of natural peptides (Robertson and Schaltze, 1970).
Careful selection of the wavelength and bandwidth of the excitation and detection systems
can optimize signal-to-noise and aid in minimizing unwanted background (Chroma, 1994).
Some fluorochromes are considerably brighter than others when localizing multiple epitopes, so that it is important to adjust the filtration to compensate when using multiple fluorochromes. The use of narrow band-pass filters can help to eliminate this problem (Fig. 6.9).

Viable Fluorescence Stains

The use of fluorescent dyes has evolved into a powerful tool for assessing cell function, viability, and proliferation. Because of the greater sensitivity, current fluorescent methods can often replace colorimetric assays for toxicity studies and high-throughput drug

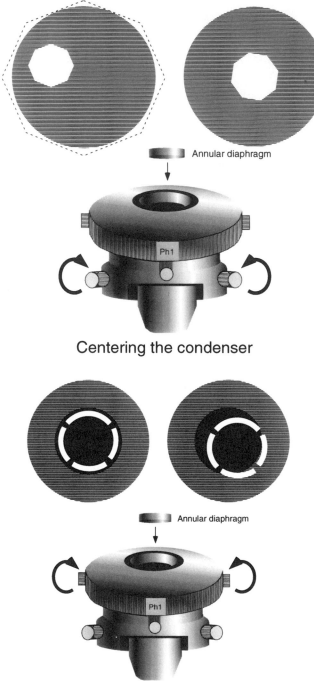

Figure 6.8. Adjusting the phase. (A) Diagram of how to adjust the phase rings. (B) View of the diaphragm. (C) Centering the annular rings. See text for instructions on adjusting the phase.

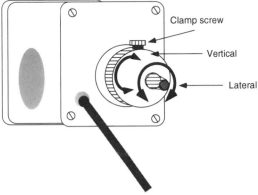

Centering the lamp

Figure 6.8. (*Continued*).

screening. Moreover, viable fluorescent tracers can be used to follow events in long-term transplantation of cells, to study dye translocation across cell membranes, to follow the movements of labeled cells in culture, and to investigate events during the cell cycle (Eidelman and Cabantchik, 1989; Johnson *et al.,* 1980). Fluorescent latex microspheres can also be used as microinjectable cell tracers as well as markers for cellular antigens and for the study of phagocytotic processes.

Dil and DiO (Molecular probes, D-282, D-275) are lipophilic dyes that have been used extensively for tracing the movement and integration of neuronal cells both *in vitro* and *in vivo* (Fig. 6.10). Cells seem to tolerate high concentrations of these dyes and remain viable for extended passages over time in culture (Kuffler, 1990; Serbedzija *et al.,* 1990).

Fluorescent cell linker compounds are also available from a number of vendors (i.e., PKH26, a red fluorescent cell linker from Sigma). These compounds have been found to be useful for cell labeling and tracking both *in vitro* and *in vivo* by incorporating aliphatic reporter molecules into the cell membrane by selective partitioning (Horan and Slezak, 1989). An esterase substrate, such as calcein, can also be used as a live cell stain, as it is retained within the cell for up to 3 hr at 37°C.

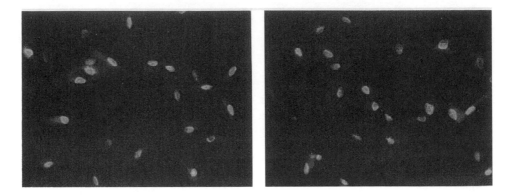

Figure 6.9. Fluorescent double stain of human Schwann cells in secondary cultures. The cell nuclei were stained with propidium iodide, which fluoresces red. The cell cytoplasm was stained for GAP (left) or S100 (right) using a green fluorescent-tagged antibody. Yellow nuclei are cells that stain both red and green. Any cells that do not stain with the specific markers will retain red nuclei.

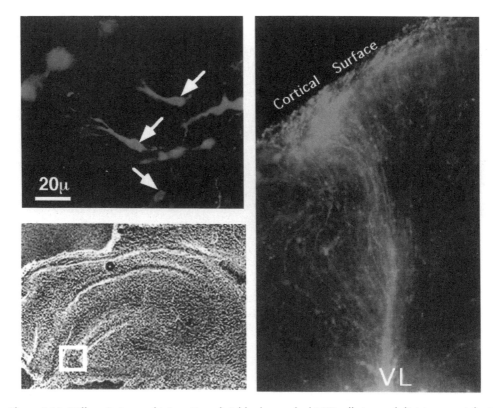

Figure 6.10. Differentiation and integration of viable dye-marked NEP cells (upper left) into neonatal rat brain. Cultured cells were labeled with the vital fluorescent dye DiI and injected near the cortical surface in newborn rat brains in the area shown in the lower left brain section. After 2 days, the animals were killed and the brain sectioned and observed with fluorescent microscopy. The NEP cells have migrated into the brain and differentiated into cells resembling cortical neurons (seen at low power on the right and higher power on the upper left).

Materials

1. Cell cultures, 2×10^7 cells/sample
2. Growth medium, serum free
3. Growth medium (serum free) with 0.1% BSA
4. 15-ml conical polypropylene centrifuge tubes
5. Clinical centrifuge
6. Microscope with epifluorescence
7. Pkh 26 kit (sigma)

Procedure

1. Trypsinize and neutralize cell suspension.
2. Count the cells by hemacytometer or electronic counter.
3. Resuspend 2×10^7 cells ($2\times$ stock) in serum-free medium and centrifuge at low speed (400 xg) for 5 min to obtain a loose pellet.
4. Prepare $4 \times 10 - 6M$ dye solution with the diluent supplied with the kit (this is a $2\times$ stock concentration). Add 1 ml to a 15-ml conical polypropylene tube and set aside.
5. Aspirate the supernatant from the pellet, allowing no more than 25 µl to remain on the pellet.
6. Tap the pellet in the remaining supernatant to loosen the cells.
7. Add 1 ml of the diluent to the cells, gently resuspend, and add this 1 ml to the tube that has the 1-ml dye solution.
8. Gently resuspend the sample.
9. Incubate at room temperature for 7 min.
10. Add 2 ml of F12.DME plus (0.1%) BSA. Allow to stand 1 min.
11. Add 4 ml of growth medium to the tube.
12. Centrifuge for 10 min at 400 xg.
13. Aspirate supernatant, transfer the pellet to a new 15-ml conical tube, and wash by centrifugation $3\times$ in 10 ml of F12/DME plus 5% BSA.
14. Resuspend cells to the desired concentration for *in vivo* injection (10^7 cells/500 µl per 100 µl) or *in vitro* assessment (10^6 cells/60-mm plate), in growth medium.

Immunohistochemistry

Immunohistochemistry can provide a rapid method for demonstrating the presence and localization of an antigen. Double labeling allows the simultaneous detection of two antigens, thus offering more information about their relative distribution. If the antigen is highly localized, it may be possible to detect as few as 1000 molecules in a cell. With advances in antibody labeling, enhancing, and particularly cell-staining techniques, immunohistochemistry can be a powerful quantitative technique as well.

There are a number of factors that influence how easily an antigen may be detected. How specific is the antibody and how diffuse is the antigen? The ideal is a large number of binding sites in a small area; but if the antigen is diffuse enough, it may be difficult to detect it from a high background signal.

The type of fixative used and its subsequent method of application are in large part empirically determined. Many fixatives alter the architectural integrity of the cell and the antibody may no longer recognize its epitope. Using frozen sections causes minimal alteration of the antigen but at the cost of poor preservation of cellular morphology. Sensitivity may further be influenced by the detection method used. Fluorochrome-labeled antibodies are excellent for visualizing subcellular components at high magnification and resolution, while enzyme-labeled antibodies offer higher sensitivity.

Polyclonal, monoclonal, and pooled monoclonal antibody preparations are generally used. Polyclonal serum will yield a strong signal, but requires careful titration to avoid an unacceptably high background. This background can be lowered by preabsorbing the serum with a saturating amount of competitor protein, either in the form of nonspecific proteins provided by BSA or serum from the antibody host. The purity and specificity of monoclonal antibodies results in a low background over a wide spectrum of concentrations. Generally, monoclonal antibodies work well with cells fixed in organic solvents or paraformaldehyde. The monoclonal antibodies that are used should be from tissue culture supernatants or purified from tissue culture supernatants or ascites fluid. Using whole ascites fluid, a not uncommon source for monoclonal antibodies, will yield unacceptably high levels of contaminating antibodies. Further amplification of the signal may be obtained by using a biotin–avidin amplification system. In these instances, one molecule (e.g., biotin) is bound to the antibody, which is bound to the antigen. Then, several molecules (e.g., avidin) linked to the detection molecule (e.g., fluorescein) bind to each biotin, thus amplifying the signal.

Cells to be used for fixation and staining can be grown on coverslips, in chamber slides, or tissue culture dishes. If chamber slides are used, they should be glass slides and should be coated with a matrix to which the cells will adhere. This is also true of the glass coverslips. Plastic tissue-culture-treated slides and coverslips can be used if the cells are not to be used with a fluorescent antibody. If glass coverslips are used, they should be rinsed and stored in acetone–ethanol. They can then be flame dried or rinsed in PBS or medium just prior to use. Suspension cultures can be grown in the usual manner and centrifuged onto slides using a centrifuge like the "cytospin" designed for this purpose.

CELL PREPARATION, FIXATION, AND ANTIBODY BINDING

Materials

1. Chamber slides, 8-well, or coverslips
2. Fibronectin (Bovine, Sigma, catalogue No. F1411)
3. Growth medium
4. 4% paraformaldehyde solution
5. 0.1 M glycine
6. PBS
7. 1% Triton-X in PBS
8. 3% BSA in PBS
9. Tris-buffered saline
10. Tween-20
11. Tris buffer

Procedure

1. Trypsinize and neutralize the trypsin in the cell suspension.
2. Seed the chambers with a cell density that will give good growth to 50% confluency within 48–72 hr. The idea is to avoid a tightly packed, confluent chamber.

3. When the cells are growing well but only semiconfluent, remove the medium and wash 1× with serum free F12–DME.
4. Mix an equal volume of F12–DME and 4% paraformaldehyde (for a final concentration of 2%). Add to each chamber well and leave for 20 min at room temperature.
5. Wash 1× with PBS and add 0.1 M glycine for 20 min.
6. Wash with PBS 2×.
7. Permeabilize with 1% Triton-×100 in PBS for 6 min. at room temperature.
8. Rinse 2× in PBS (leave a coating a detergent with each wash).
9. Remove all but 200μl PBS. Apply the first antibody, 200 μl/well (1:25 if dilution is unknown).
10. Incubate for 30 hr at 37°C.
11. Wash 4× in PBS plus 0. 1% Tween-20 (5 min./rinse). Leave 200 μl in well each rinse.
12. Add the secondary antibody in 3% BSA–PBS or 10% host serum.
13. Repeat steps 9–11.
14. If using an alkaline phosphatase detection method, wash in Tris buffer for 5 min.

Detection

The detection step uses either enzyme-labeled or fluorescent-labeled reagents. The enzyme label is detected using a chromogenic substrate that precipitates subsequent to the enzyme reaction. This results in an insoluble colored product at the site of localization of the antigen. Secondary antibodies conjugated to either horseradish peroxidase or alkaline phosphatase are commercially available, and when used with a wide range of substrates can produce several different colored products that vary in intensity. The use of a variety of substrates may be required, depending on the level of background and sensitivity, and when two colors are required. For horseradish peroxidase, the most common substrate is diaminobenzidine (DAB), which yields an intense dark bronze or brown color. If DAB yields too high a background, chloronaphthol (blue-black) or aminoethylcarbazole (AEC), which gives a less intense red color, can be used. For alkaline phosphatase, the most common chromagen is bromochloroindolyl phosphate–nitro blue tetrazolium (BCIP–NBT), which yields a purple-black color. Naphthol-AS-Bl-phosphate–new fuchsin (NABP–NF) or fast red is also frequently used. These colored products produce a permanent record and the slides can be stored and reviewed after long periods of time, whereas fluorescent probes fade with time and some are readily quenched during the process of viewing the slides.

Materials for Horseradish Peroxidase-Conjugated Second Antibodies

1. 0.05M Tris buffer pH 7.6
2. DAB tablets (Sigma catalogue No. 5905)
3. Hydrogen peroxide 30% (stored at 4°C for up to 1 month)
4. Whatman No. 1 filter paper

Procedure

1. Dissolve one DAB tablet in 15 ml Tris buffer (0.05 M, pH 7.6).
2. Add 0.1 ml of a 3% solution of hydrogen peroxide. Filter the precipitate if necessary.
3. Add solution to the well sufficient to cover the cells.

4. Incubate at room temperature for up to 20 min. Stop the reaction by washing with water.
5. Mount with 50% glycerol or Permount.

Materials for Alkaline Phosphatase-Conjugated Second Antibodies

1. Naphthol AS-Bl-phosphate
2. New fuchsin (Sigma catalogue no. N0638)
3. Sodium nitrate
4. Naphthol AS-TR (Sigma catalogue no. N5875)
5. 0.2 M Tris buffer (pH 9.6)
6. 2 N HCl
7. Dimethylformamide
8. 20 mM EDTA

Procedure

1. Dissolve 1 mg of new fuchsin in 0.25 ml 2N HCl.
2. Dissolve 1 mg sodium nitrate in 0.25 ml H_2O.
3. Dissolve 10 mg naphthol AS-TR in 0.2 ml Dimethylformamide.
4. Add the new fuchsin solution to the sodium nitrate solution and mix for 1 min. Add this to 40 ml of 0.2 M Tris (pH 9.6) buffer.
5. Add naphthol AS-TR solution to the solution in step 4.
6. Cover the well with the solution and leave for up to 30 min at ambient temperature. Stop the reaction with 20 mM EDTA.
7. Mount with 50% glycerol or Permount.

BRIGHT FIELD

Generally, bright-field microscopy is used in the observation of specimens that are fixed and stained, such as histology sections. The microscope used is usually upright as opposed to inverted. This type of conventional illumination will not reveal brightness differences between the structural details in the living cell and its surroundings because the image lacks contrast, but it is ideally suited for pathology and histological applications in which the morphology and immunohistochemical staining are of paramount importance in preserved specimens. Bright-field microscopy is of little use for viewing living tissue culture plates, but it may prove useful to have a second bright-field microscope if the work in the laboratory depends a lot on stained tissue or cultures or uses immunohistochemical staining to gather data.

DARK FIELD

In dark-field microscopy, the illuminated object is bright against a dark background. This is achieved by means of a central solid spot in the condenser that allows illumination of the cell or particle only from the side. The extremely high contrast provided by this type of illumination makes it possible to visualize very small particles in the cell that are beyond the resolution of the phase contrast microscope.

Dark-field microscopy is a method that is useful when trying to detect minute particles. Examples might include autoradiography or *in situ* hybridization in tissue sections or cultures grown on slides and processed. The DIC condenser used for dark-field illumination creates a cone of light in which the direct or oblique rays do not enter the objective. When these rays intercept a particle, however, they are scattered into the objective, causing the particle to appear luminescent against a black background (Fig. 6.11). Dark field can best be viewed with the smaller magnification objectives, that is, 8× or 10×. A 20× objective can be used, but it is preferable to use one that does not have high light gathering capability (one with a low NA), to avoid an undesirable high background. Special objectives for dark-field use are available if higher magnifications are needed.

A dark-field format can be achieved through the use of a phase contrast condenser and bright-field (nonphase) objectives. The phase ring in the condenser should be larger than the objective used. This will then produce a dark-field cone of light where the direct ray of illumination will not enter the objective. Usually, the use of a 10× objective and a ph3 annular ring will be sufficient. There are also after-market sources for obliquely lit slide holders, which most popular microscope stages can accommodate, that create a dark-field effect while allowing the additional use of filters on the microscope itself for further image enhancement.

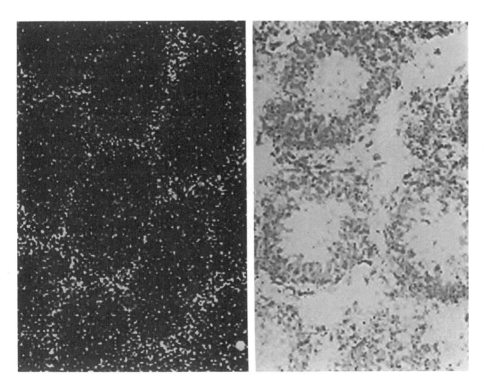

Figure 6.11. A frozen section of rat testis treated with autoradiography after [^{125}I]inhibin has been bound, and then the section was counterstained. The left panel shows a bright-field view that allows one to see cellular morphology, while the dark-field view of the same section on the right emphasizes the silver grains.

ADDING THE THIRD DIMENSION

A vexing problem in conventional microscopy has been that objectives with a numerical aperture high enough to resolve fine structure will be limited by an extremely shallow depth of field. This restricts the information gathered to two dimensions in which the complex structural relationship of the subcellular components is difficult, if not impossible, to discern. Even DIC illumination, which provides a surface view, is two-dimensional and suffers from a shallow depth of field. Microscopes that allow three-dimensional viewing of specimens on slides are relatively new. Scanning confocal microscopes utilize a computerized reconstruction of data viewed through sections along the z-axis, thereby generating a three-dimensional view (Boyde, 1985, 1986). The Edge high-definition stereo light microscope (Edge Scientific Instrument Co., Santa Monica, CA), as seen in Fig. 6.12, is set up to allow real-time three-dimensional viewing in phase, bright field, dark field, or fluorescence. Its illumination system supplies distinct left and right images for true stereoscopic observation as well as three-dimensional and conventional photomicroscopy. As data acquisition and analysis software become available, this combined technology will become a powerful tool for investigators needing to localize, analyze, and quantitate intracellular information in real time *in vitro*.

CONFOCAL MICROSCOPY

Confocal microscopy, particularly when it is used in the scanning imaging mode, has advantages over conventional instruments, particularly for three-dimensional imaging of thick (50–200 μ) specimens. Higher resolution, contrast, and depth of field offer added information as to the cell's physiology, motility, and three-dimensional structure. A conventional microscope utilizes a large-area incoherent illumination source via the condenser and relies on the objective for resolution. The confocal microscope utilizes a point source and a point detector created by placing a pinhole in front of the detector (Wilson, 1990). This results in increased image resolution, as both the light source and the photomultiplier are taking part in the image-forming process. A finely focused spot of light, coupled with electronic image processing, allows for multiple images to be scanned throughout the specimen and then stored and reconstructed, utilizing a wide variety of image-enhancing techniques. Applications for this type of imaging technology include cellular localization of macro-

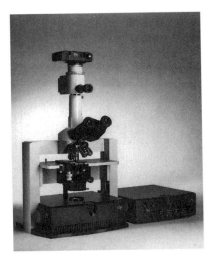

Figure 6.12. The Edge three-dimensional microscope provides real-time three-dimensional viewing of live or fixed specimens. (Photo courtesy of Edge Scientific Instrument Corporation)

molecules and cellular structures (cytoskeletal filaments and organelles), localizing a particular cell type in a tissue, multiple image labeling, and imaging of three-dimensional structures over time, referred to as four-dimensional imaging (Stricker *et al.*, 1990).

ADDING THE FOURTH DIMENSION

There is even more information to be gathered from cells if the dimension of time is added. Cells are living organisms that exhibit many characteristic phenomenon that can be studied if one has the capability to observe and record the movement of cells. Cells may move for locomotion of individual cells, for example, the swimming movement of a sperm cell. They may be induced to move in order to avoid or to be attracted to another cell, surface, or chemical compound. Parts of cells such as flagella or cilia may move. The surface of the cells may move in response to growth factors or insults, leading to changes in cell shape or ruffling or blebbling of cell membranes. Apoptotic and necrotic cell death are each accompanied by distinct movements of the cell surface. The type of movement will determine the rate of photography and the type of equipment required to contain the cells during filming.

REAL-TIME VIDEO

Real-time video can be accomplished with a VCR, monitor, and a high-resolution black and white or color charged coupled device (CDC) video camera. Some of these cameras are cooled down to $-37°C$ to reduce electronic noise and produce a high-resolution image at extremely low light levels. Even if your video aspirations are modest, the purchase of a super-VHS VCR rather than a regular-VHS VCR is recommended, as the increase in resolution far outweighs the difference in price. Most consumer super-VHS VCRs are adequate for this purpose, and usually have editing capabilities as well. A high-resolution monitor (8″ or 13″) with SVHC or component input will afford high-resolution viewing.

TIME-LAPSE VIDEO

Time-lapse cinematography has been available for many years. Recent advances in time-lapse VCRs offer the investigator an affordable way of utilizing this technology in a video format. Videotape is relatively inexpensive and comes in a wide range of lengths, usually from 10 min to the standard 120 min in VHS and super-VHS.

An alternative to the time-lapse VCR is a video capture board that can be used on either Macintosh or Intel platforms. When used with appropriate software, these boards can capture images via the video camera at specified intervals and can be played back directly on the computer. This does not capture continuously, as with a VCR; but for many applications, a continuous video may not be necessary. In addition, with videotape recording, postprocessing of the images may be required. Fifty or 100 hours of culture captured on videotape usually needs to be edited down to 3–5 min for final presentation. This requires an additional investment in video-editing equipment.

Regardless of whether one chooses a VCR or capture board, if long-term culture is to be done on the microscope stage, an appropriate physical environment is required. Commercial incubator chambers are available for microscopes. They can be miniaturized as in the Leiden chamber or they can cover the entire microscope stage (Fig. 6.13). Incubator

Figure 6.13. An incubator box can be constructed to fit over the entire microscope and provide temperature and CO_2 control for time-lapse microcinematography.

chambers can also be fabricated to the investigator's specifications either in-house or by an outside plastics fabricator, usually more economically than purchasing a prefabricated unit. Maintaining an even distribution of heat over the surface of the stage is critical. We utilize two systems. The modified in-house can maintain a constant temperature and can be left running continuously (Fig. 6.14). These use temperature controllers with miniature RTD or thermocouple probes. CO_2 can be supplied from an in-house source or from gas mixtures of 21% O_2, 74% NO_2, and 5% CO_2 supplied commercially, in size T cylinders equipped with two-stage regulators. In either case, the gas is first passed through a phenol red solution (15 mg/liter phenol red, 1.2 g/liter $NaHCO_3$ in water) as an indicator of pH, and a copper sulfate solution (1 g/liter in water) to decontaminate and aid in humidifying the gas. The gas then enters a 65-mm flowmeter (1 liter/min), which is adjusted to a rate of 300–500 ml/min. This flow rate will provide an adequate turnover and sufficient humidity for short-term (48–72 hr) cultures. A small dish of water can be placed in the chamber for additional humidity if long-term culture is required. An acrylic chamber with an optically clear glass top is used to house the culture dish and aids in containing the gas–air mixture. Further details on these systems are supplied in Appendix 2.

Figure 6.15 is a low magnification phase contrast film of the association of cells into "balls." This film was done with a 16-mm movie camera and the prints were made from single frames. Figure 6.16 shows cells dying by necrosis on the top and by apoptosis on the bottom. These frames are taken from time-lapse videos of the two cultures recorded directly onto videotape. The times at which the individual frames were taken are displayed directly on the tape. One can often convey the information learned using time lapse by such composite prints for publication. However, the impact of seeing a cell culture as a living, dynamic interactive organism is best conveyed by viewing the film or video. One hopes that the increased use of electronic publishing and the Internet will allow some video data to be available over the Internet to complement published still photographs.

Videotapes of cells in culture can be rented from the Society for In Vitro Biology. Videotapes have also been produced under the auspices of the American Society for Cell Biology and are available for purchase (Fink, 1991, 1995; Pickett-Heaps and Pickett-Heaps, 1996).

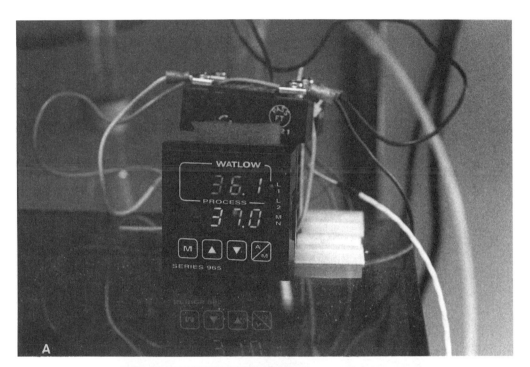

Figure 6.14. (A, B) Parts of a temperature control system for an incubator box.

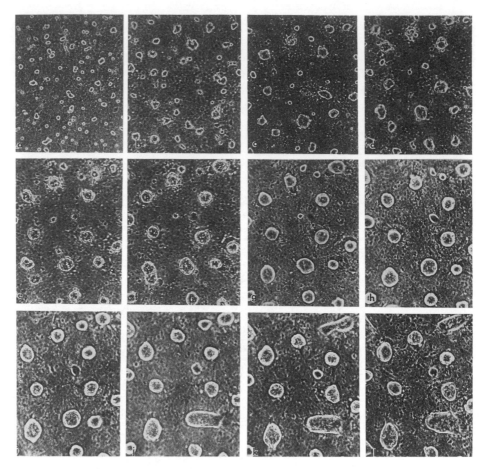

Figure 6.15. A series of frames from time-lapse photography of the coculture of TM4 Sertoli cells plated on a monolayer of TR-M peritubular myoid cells. The TM4 cells form "balls" of cells only when plated on a TR-M monolayer.

HIGH-SPEED VIDEO

Since cells usually move slower than we do, high-speed video is not widely used in cell culture. It has been used, however, to slow down and analyze the movement of mammalian sperm, cilia on mammalian cells, and the movement of microorganisms that depend on cilia or flagella for locomotion. This is basically a video taken at high speed and slowed down for viewing.

———— LOOKING MORE CLOSELY ————

All of the above methods of looking at cells are limited by the wavelength of visible light. If one wishes to view smaller objects, electrons are substituted for visible light and the "image" must be translated from electrons to an image we can interpret visually. Both

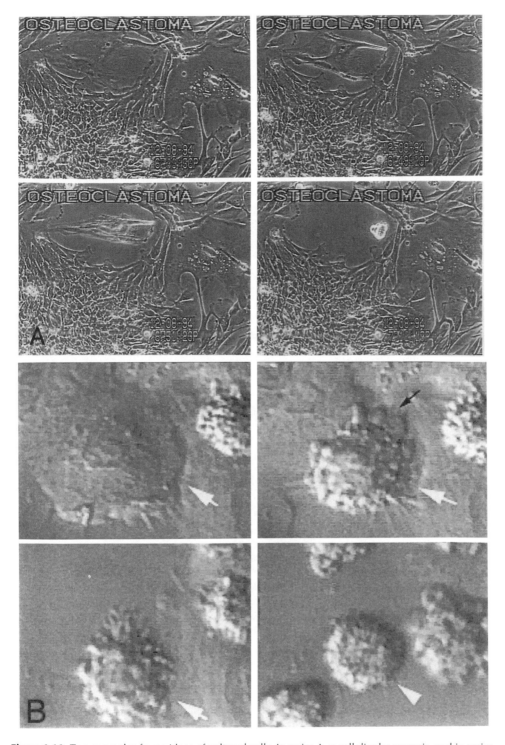

Figure 6.16. Two examples from videos of cultured cells. In series A, a cell dies by necrosis, and in series B, one can see a number of cells dying by apoptosis. Note the blebbing in the apoptotic cells.

of the microscopes discussed can look at only fixed and therefore dead material and "see" only in black and white. However, one can look much more closely at the cells' morphology. The book by Fawcett (1981) is an excellent atlas of cell morphology at the electron microscopic level.

SCANNING ELECTRON MICROSCOPY

The scanning electron microscope (SEM) is a valuable tool for looking at the surface of cells at high magnifications. To prepare cells for scanning electron microscopy, the cells are fixed and dehydrated and the surface is coated with a reflective metal such as gold or platinum. The image is then bombarded with an electron beam. The image seen in the microscope is created from the reflected electrons. "Shadows" seen in the photographs give the images a three-dimensional feel. As can be seen in Figs. 6.3 to 6.5, cell lines that have a similar morphology when viewed in the phase microscope can have quite distinct morphologies when seen in SEM. Other scanning electron micrographs can be seen in the figures in Chapter 7 and other chapters. Surface features such as microvilli, cell–cell association, and cell shape (e.g., thick or flat) can only be seen by scanning electron microscopy. As discussed in Chapter 7, the electron microscope is also required to visualize mycoplasma and viruses.

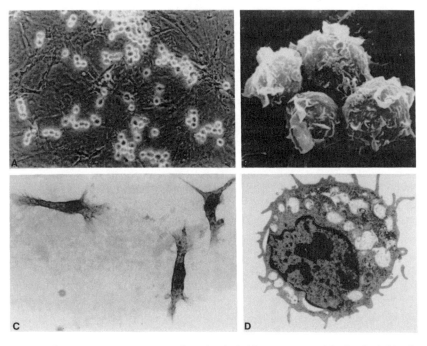

Figure 6.17. (A) Phase, (B) SEM, (C) TEM, and (D) bright-field image (stained for bright field) of testicular macrophages in primary culture. The different types of microscopy provide different information to the researcher.

The transmission electron microscope (TEM) can visualize subcellular components such as mitochondria, golgi, vesicles, and other organelles. Under special conditions one can even see individual macromolecules such as DNA and myosin–actin interactions. This, then, is the ultimate in looking closely. However, since one is looking at only a fraction of the cells or of an individual cell at any given time, finding a rare event to look at may be

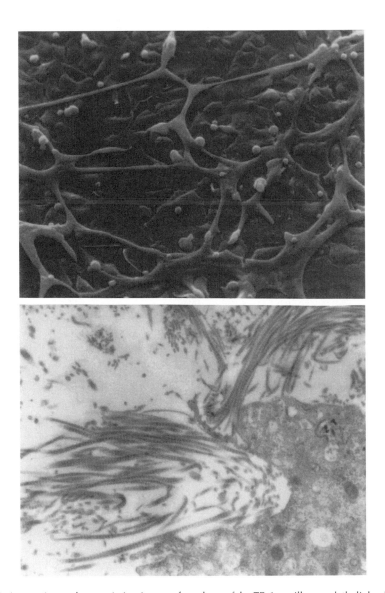

Figure 6.18. A scanning and transmission image of a culture of the TR-1 capillary endothelial cell line that has formed tubulelike structures *in vitro*. (A) The SEM view allows one to see the network of tubules, while (B) the higher magnification section through one of the tubules shows that it is hollow, with collagen and matrix material inside.

difficult. Additionally, since the cells must be fixed, dehydrated, chemically modified (e.g., osmium treatment), and sectioned to be seen, there is ample opportunity for the introduction of artifacts in the process of preparing the cells. Here, there is nothing like an experts' help. If you are unfamiliar with looking at cells in an electron microscope, find someone who has done so. Talk to them first about their experience with the best method to grow and fix cells for observation. Look at some of their work to familiarize yourself with how cells look in section and what the various common artifacts of fixation look like. If you are working with someone else who will be doing the microscopy, familiarize them with the objectives and the protocols used in the experiment. If possible, show them the cells before fixation and point out any features that you want to observe more closely. In this instance it is important that the electron microscope operator know what he or she is looking for. A hundred photos of healthy quiescent cells when you are only interested in looking at dividing cells are not much help—and a terrible waste of time for the microscopist.

Figures 6.16 and 6.18 illustrate the levels of information that can be obtained by using different ways of looking at cells. Figures 6.17 and 6.18 show several different views of both cell line and primary cell cultures. Clearly, more detail can be seen by looking at the TEM, which shows the actual cell organelles. However, the SEM picture gives a better feeling for the cell shape and surface characteristics, such as ruffles, and the phase photograph is the only one that gives an instant impression of how many cells in the culture are undergoing mitosis. Figure 6.18 shows SEM and TEM views of "tubes" formed in vitro by a capillary endothelial cell line isolated from rat testis. Again, the overall pattern of the branching structures is best shown in the SEM picture. Only the TEM picture shows that the tubes are hollow and the basement membrane material (identifiable as collagen from the TEM) is vectorially secreted to the inside of the tube. Time-lapse movies of such cultures added the surprising bit of information that these structures are not static, but rather have some cells moving along the tubes at all times. Thus, no one method of viewing cells is "right," but rather each gives different types of information and is best for providing answers to different types of questions.

REFERENCES

Boyde, A., 1985, Stereoscopic images in confocal (tandem scanning) microscopy, *Science* **230:**1270–1272.

Boyde, A., 1986, Applications of tandem scanning reflected light microscopy and three-dimensional imaging, *Ann. NY Acad. Sci.* **483:**428–439.

Chroma, T. C., 1994, *Chroma Handbook of Optical Filters for Fluorescence Microscopy,*

Eidelman, O., and Cabantchik, Z. I., 1989, Continuous monitoring of transport by fluorescence on cells and vesicles, *Biochim. Biophys. Acta* **988:**319–334.

Fawcett, D., 1981, *The Cell,* 2nd ed. Philadelphia: W. B. Saunders.

Fink, R., 1991, *A Dozen Eggs: Time Lapse Microscopy of Normal Development,* Sunderland, MA: Sinauer Associates.

Fink, R., 1995, *CELLabration,* Sunderland, MA: Sinauer Associates.

Horan, P. K., and Slezak, S. E., 1989, Stable cell membrane labelling, *Nature* **340:**167–168.

Johnson, L. V., Walsh, M. L., and Chen, L. B., 1980, Localization of mitochondria in living cells with rhodamine 123, *Proc. Natl. Acad. Sci. USA* **77:**990–994.

Kuffler, D. P., 1990, Long-term survival and sprouting in culture by motoneurons isolated from the spinal cord of adult frogs, *J. Comp. Neurol.* **302:**720–738.

Pickett-Heaps, J., and Pickett-Heaps, J., 1996, *Living Cells: Structure and Diversity,* Sunderland, MA: Sinauer Associates.

Robertson, E. A., and Schaltze, R. L., 1970, The impurities in commercial glutaraldehyde and their effect on the fixation of brain, *J. Ultrastr. Res.* **30:**275–287.

Serbedzija, D. N., Frazer, S. E., and Bonner-Fraser, M., 1990, Pathways of trunk crest cell migration in the mouse embryos as revealed by vital dye labeling, *Development* **108:**605–612.

Stricker, S. A., Paddock, S., and Schatten, G., 1990, Laser scanning confocal microscopy of living sea urchin embryos: 3-D reconstruction and calcium ion imaging, in: *The American Society for Cell Biology Thirtieth Annual Meeting,* Vol. III (T. Wilson, ed.), The Rockefeller University Press, San Diego, CA, pp. 113a.

Wilson, T. (ed.), 1990, *Confocal Microscopy,* Academic Press, New York.

CHAPTER 7

Contamination: How to Avoid It, Recognize It, and Get Rid of It

One occurrence that every person who tries to grow mammalian cells *in vitro* has to deal with sooner or later is contamination. As a problem, it can vary from irritating to catastrophic. The best solution? Avoid getting contamination in the first place. Failing this, the next best thing is to destroy all the contaminated cultures. However, since neither solution is likely to work all of the time, in this chapter we shall discuss how to recognize contaminated cultures and what to do about them when you do get contamination. One approach to contamination you cannot do is ignore it. Contamination in a cell culture will influence virtually any parameter you might wish to study, even if it does not immediately kill the cells. Do not ever use contaminated cultures to get numbers (we will not call it data) on the grounds that "they are just a little contaminated" or "the cells are still alive." Any numbers you get will be misleading rather than helpful, and a waste of time.

STRINGS, WIGGLIES, AND PRETTY BALLS OF FLUFF

The easiest contaminants to deal with are those you can see without any doubt. Some, like mold, can be seen even without a microscope (Fig. 7.1). Others, like bacteria and yeast, are too small to be seen except under relatively high power in the microscope (Fig. 7.2), but they turn the medium visibly cloudy and frequently drive the medium pH to very acidic levels with their metabolic by-products. Thus, any culture that seems to be unusually acid, looks cloudy, or has fluffy balls or strings in it should be removed from the incubator and observed in the microscope. *Do not open the culture dish.* Remember that all contaminated cultures are potential sources of contamination for other cultures and can easily aerosolize.

If the microscopic examination of the cultures confirms a contaminant, tape the tissue culture dish closed or place it in a sealed bag to be autoclaved. Immediately wipe down the

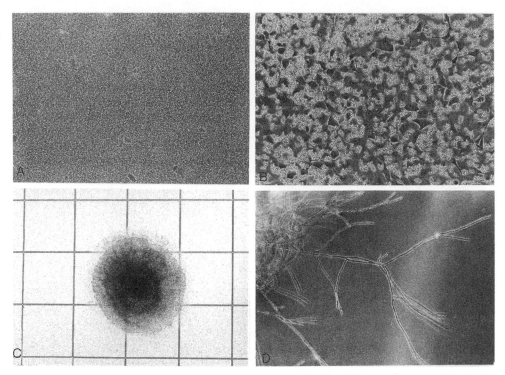

Figure 7.1. Mold, yeast, and bacterial contamination in mammalian cell culture viewed through the phase contrast microscope. This is what you see when you look at the cultures in the microscope. Panels A, B, and C are viewed at 300× magnification and panel C at 60×. The culture in A contains bacteria and in B, yeast. The mold colony in C is growing over the grid of a 150-mm culture dish. The edge of the same colony is shown in panel D.

microscope stage, bench top, and your hands with a disinfectant such as alcohol. Mold can grow on the outside of the culture dishes and on the plastic, metal, or glass surfaces in the laboratory. Labeling tape and paper labels are particularly susceptible to growth of mold. If there are many contaminated dishes, they can be placed in a separate autoclave bag and immediately removed from the room for sterilization.

Yeast, mold, and bacterial spores are ubiquitous. They are found adhering to dust and water vapor in the air and to hair, skin, clothing, and shoes. While filtering the air entering the culture room through HEPA filters will help a great deal in keeping down contaminants, unless each person entering the room goes through a thorough shower and gowning to avoid contamination (something that is not done for a normal culture room), the bugs will catch a ride. Some of the worst instances of yeast contamination we have experienced have been traced to laboratory personnel who were baking bread or brewing beer at home. They were bringing their home to work rather than taking their work home. Therefore, the best safeguard is to avoid unnecessary traffic into the room where the culture work is done. A "sticky mat" placed inside the entrance door to the culture room will help remove dust and spores from the shoes of people entering the room. The paper must be changed at least daily to be effective. Laboratory coats that are to be worn in the culture room only and left hanging near the door when the personnel leave should be available. Bacterial, mold, and fungal contaminants are the most common because they are designed by nature to be resistant to

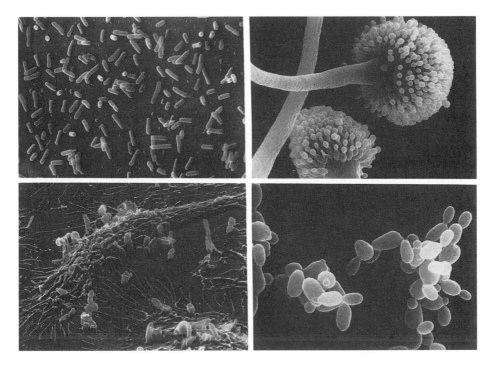

Figure 7.2. Scanning electron micrographs of contamination in cell culture. SEM allows one to see more detail of these small organisms: (A) Bacterial contamination; (B) sporulating mold; (C) pseudomonas on cells; and (D) budding yeast.

desiccation and to remain for long periods of time in a dormant state but to immediately replicate when they find a good food source, such as cell culture medium.

Food and drink should not be brought into the culture room. If it is possible to do so, locate the sink adjacent to the culture area entryway so that anyone working directly with the cultures or handling plates can wash their hands prior to entering the culture area.

THINGS YOU CANNOT SEE
CAN HURT YOU

MYCOPLASMA

Mycoplasma contaminants grow quite slowly, compared to bacteria and yeast, and do not immediately or directly destroy the cells. They are known to alter the function and metabolism of the cell culture, cause chromosomal aberrations, affect cell surface antigenicity, interfere with nucleic acid synthesis, and generally change cell behavior (Kotani *et al.,* 1987; McGarrity, 1977; Van Diggelen *et al.,* 1977a, b). Mycoplasms are too small to be seen directly except by fluorescence labeling and the use of very high power objectives or electron microscopy (Phillips, 1977, 1978), but any significant change in the growth and function of the cultures should alert the investigator to the possibility of a mycoplasma contamination. In our laboratory, we mycoplasma test any culture coming into the laboratory and any stock to be frozen down. We also routinely test our cell lines every 6–12 months. If contamination is detected, cells should be tested more frequently until it is clear that the contamination has been eliminated. In addition, we maintain our cultures in antibiotic-free medium. While this does not preclude the introduction of adventitious agents such as mycoplasma, it serves two purposes: (1) It requires exceptionally careful handling and manipulation of the cultures to avoid overt contamination; and (2) infections by other microorganisms will serve as a warning that there is an observable problem that needs to be dealt with, hopefully before the introduction of a mycoplasma.

There are fluorescent staining methods for detecting mycoplasma that exploit the ability of Hoechst 33258 dye to bind specifically to DNA. Mycoplasmas contain DNA and can be seen as bright, punctate staining in the cytoplasm (Fig. 7.3). This is a good routine test for screening purposes. Scanning and transmission electron micrographs of mycoplasma infected cells are shown in Fig. 7.4. These organisms are quite varied in morphology and much smaller than the host cells. Commercial kits are available for mycoplasma screening

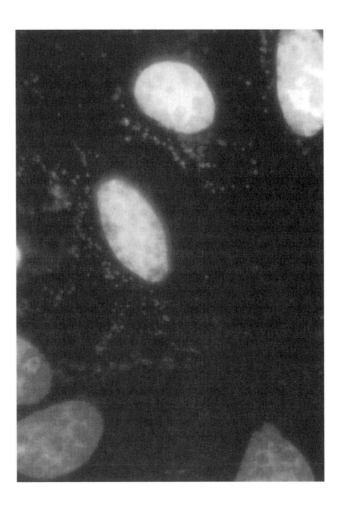

Figure 7.3. Mycoplasma-contaminated cells. Stained with Hoeschst stain. (Photograph courtesy of Dr. Gerald Massover)

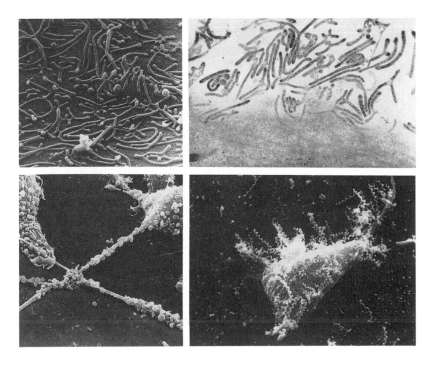

Figure 7.4. Electron micrographs of mycoplasma growing on cell cultures. Mycoplasma, the smallest free-living organisms, are too small to be seen with a light microscope, although fluorescent staining allows one to see them at high power (previous figure). (A) SEM and (B) TEM of *Mycoplasma pneumonia* on mammalian cells. The mycoplasma are about the same size as microvilli, but can be distinguished by being able to see both ends. (C) a more bulbous-shaped mycoplasma and (D) spiral spiroplasma growing on insect cells.

that depends on cultivation of mycoplasma or detection of mycoplasma using specific antibodies or polymerase chain reaction (PCR) (for example, the "Panverra" kit for PCR). Out of 3409 samples screened in one facility, 3.7% were positive by culture and an additional 0.7% positive by both Hoechst stain and PCR. No contamination was detected by culture that was not also positive using PCR and DNA staining detection methods (M. Roy, unpublished data). It would therefore seem advisable to use DNA staining or PCR for mycoplasma detection if the equipment necessary for these techniques is available.

METHOD FOR FLUORESCENT DETECTION OF MYCOPLASMA

Materials

1. Hoechst 33258, 1 mg/ml stock solution (protect from light at 4°C)
2. Acidic acid:methanol (1:3) fixative
3. Mounting solution (50% glycerol in water)
4. Nunc two-well or four-well chamber slides
5. PBS
6. Microscope with UV light source and 100× oil immersion objective suitable for use with fluorescence (*Note:* Some lenses will not transmit UV light)

Procedure

1. Seed the culture so that 50% confluence can be reached in 48–72 hr.
2. Incubate at 37°C, 5% CO_2, humidified. Place the chamber slides in a 150-mm tissue culture dish, if available, to facilitate handling and avoid spillage.
3. Remove the medium.
4. Wash 1× with PBS.
5. Add PBS:fixative (1:1); leave at room temperature, 5 min.
6. Remove PBS:fixative and add straight fixative.
7. Remove and add fresh fixative; leave at room temperature for 5 min.
8. Remove fixative and rinse with distilled or purified water.
9. Make up a working solution of 100 ng/ml Hoechst 33258 in PBS.
10. Add working solution to chamber slide and leave at room temperature 10 min.
11. Remove the dye and rinse with water.
12. Remove the upper chamber from the slide. Make sure no silicon adhesive is still attached.
13. Mount a coverslip with a drop of mounting solution and examine the slide under epifluorescence with a 360-nm excitation 490 barrier filter.

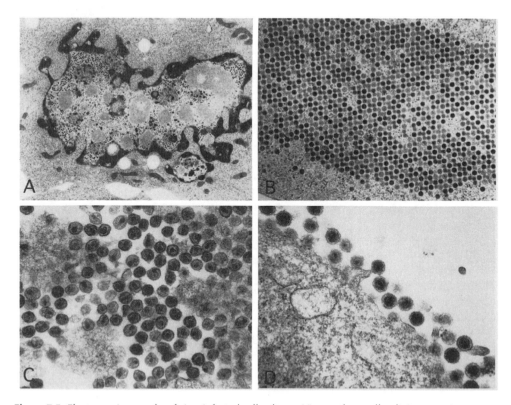

Figure 7.5. Electron micrographs of virus infected cell cultures. Viruses, the smallest living organisms, must replicate inside cells, using part of the cell's protein synthesis molecules. They can only be seen using transmission electron micrography, but they can significantly alter cellular functions. (A, B) Adenovirus. (C) Human herpes virus 6. (D) HIV on the surface of a human T cell.

Endogenous and contaminant viruses present in cell culture systems can be difficult and expensive to identify. Representative problem viruses include Sendai virus, simian virus, murine or human hepatitis virus, Epstein–Barr virus, human cytomegalovirus, and human immunodeficiency viruses (HIVs) and other endogenous retroviruses. Protocols have been established by organizations such as the American Type Culture Collection (ATCC) and government agencies to screen for latent and chronic viruses. Additional protocols have been developed where cell lines are to be used for production of biological pharmaceuticals (Lubiniecki, 1990) (see Chapter 12). Most exogenous retroviruses are undetectable in cell cultures, except by serological or biochemical methods, or by the use of electron microscopy. Several electron micrographs of virally contaminated cells are shown in Fig. 7.5. Infected cultures do not necessarily display overt morphological change.

In general, all human cells should be handled as if they were a biohazard. Cells should be handled in a contained hood with air HEPA-filtered going into and out of the hood. All cell culture waste should be autoclaved after use and spent media should be treated with a viricide and disposed of appropriately. If primary tissues are needed, they can be tested for hepatitis and HIV when they are removed from the donor. Unless the laboratory is equipped for handling these agents, such infected tissues and cell lines should not be used, since viruses from cells and tissues can infect human handlers. Likewise, mouse viruses can spread from cell cultures to an animal colony. Therefore, many animal facilities require viral tests of cultures before the cells are injected into animals in a colony. If primary cultures are to be studied or used to establish new cell lines, always obtain animals from a reliable source that tests for viruses and mycoplasma in the animal colonies. If human sera are to be used in culture, use only sera that have been screened for viral contaminants.

——— CROSS-CULTURE CONTAMINATION ———

The most difficult type of contamination to detect may well be contamination of one cell type by another cell line normally carried in the laboratory. This type of contamination can be carried for years without detection. It clearly will invalidate all publications arising from the use of such cross-contaminated cell lines. Hukku *et al.* (1984) report the incidence of cross-contamination of cell lines to be in excess of 35%. Cross-contaminated cell lines can be carried as mixed cultures for long periods of time, or the contaminating cell might take over the entire culture, causing the loss of the original cell line. If the cells lines are similar in appearance and derived from the same species, then this situation may continue for years without detection.

When a new cell line is established in a laboratory or is brought into the laboratory from another laboratory—commercial or academic—it should be typed to determine identity. Kits are available for isozyme typing that can determine the species of origin of a given cell line. The use of antibodies can also determine the species of origin and whether the cells in a given culture arise from only one species. Karyotyping will also give information on the species of origin and the sex of the individual from which the cell culture arises. In some cases, with human cell strains we have been able to differentiate cell lines derived from the same cell type but arising from individual donors by using marker chromosomes. Karyotyping and cell identity determinations can also be done by commercial laboratories. One such report is shown in Fig. 7.6.

These procedures are somewhat expensive and therefore will probably not be repeat-

We have examined cell line HSS p3, and we have the following results to report.

Isozyme phenotypes:

LDH	G6PD	PGM1	PGM3	ESD	Me-2	AK-1	GLO-1
human	B	1	1–2	1–2	1	1	2

We calculate the phenotypic frequency of this phenotype to be 0.0057. In other words, less than 1% of cell lines might be expected to have an isozyme phenotypic profile identical to this.

Reaction with species-specific antisera:

All cells reacted with fluorescein-conjugated anti-human antiserum. None of the cells reacted with fluorescein-conjugated anti-mouse antiserum.

Cytogenetics of cell line HSS p3:

Chromosomes: human
Y chromosome: not present (by QM staining)
Chromosome count ploidy distribution/100 metaphases:
    100 metaphases with 46 chromosomes (2N = 46)
Exact chromosome counts/30 metaphases:

No. of metaphases   30
No. of chromosomes   46

Giemsa banded chromosomes: Ten karyotypes were prepared from metaphases with 46 chromosomes. Two of the karotypes are enclosed.

Figure 7.6. Copy of a report on the experiments carried out to determine cell line identity. The report details the results using karyotyping, species-specific antisera, and electrophoresis for polymorphic enzymes to determine that the cells in the culture are of human origin, with no contaminating mouse cells seen.

ed routinely (like mycoplasma testing). However, if the cell lines are all characterized when they are brought into the laboratory and a frozen cell bank laid down at that time, then one can go back to this bank if the cells begin to look or behave differently. The best protection against cross-culture contamination is good technique:

1. Do not have two different cells lines in the hood at the same time.
2. Do not ever put the same pipette tip or anything else into two sequential stock cultures.
3. Wipe down the hood surface and wash hands, or change gloves, between handling different cell lines.
4. Store frozen cells in the vapor phase of the liquid nitrogen and/or use glass freezing vials and leak test.

As with other types of contamination, if you suspect that your cultures have been cross-contaminated, destroy them. Then thaw a frozen vial, grow it up, and retest for identity. If this is impossible, the suspected mixed population can be recloned by limiting dilution and clones screened and tested. This will only work, however, if the culture is still mixed rather than replaced by the contaminating cell line.

WHAT CAN YOU DO TO PREVENT CONTAMINATION?

Prevention of contamination is mainly a matter of vigilance and the trade-off between trying to maintain a totally sterile environment in the culture room and the inconvenience, time, and cost involved in doing so. In those cases where prevention of contamination is of primary importance, then the culture room should be restricted to personnel who work in it. It should be entered through an airlock that is maintained at an air pressure greater than that of the outside and less than the culture room. The entry room should provide hair and shoe coverings and laboratory coats or overalls. The floor of the entry room should have a sticky mat to catch the dust on shoes. The walls inside the culture area should have UV lights that go on when the room is empty and HEPA-filtered air. (*Note:* UV lights do cause degradation of plastic components of equipment and paint and vinyl flooring. For safety, the lights should always automatically go off when the door is opened and should have to be reset from outside the culture room after closing the door.)

For those who do not wish this level of containment, the movement of people and equipment in and out of the room still should be minimized. People should have laboratory coats that are worn only in the culture room. The culture room should be maintained under positive air pressure and the air HEPA-filtered if possible. Disposable sticky mats just inside the culture room door are a reasonable precaution and minimize tracking dust and spores into the culture room.

The inside of the room or area used for cell culture should be well organized and not overcrowded. It is best to store cases of supplies elsewhere and have a cabinet in the room for the minimum of supplies necessary, for example, for a weeks' use. This can be restocked periodically. All tissue culture supplies should be used on a first in–first out basis, since sterility cannot be guaranteed after prolonged storage. The following tips will help maintain a clean culture room and minimize contamination:

1. Bench tops, hood surfaces, microscopes, and other work surfaces should be wiped down periodically with ethanol or another germicide.
2. The inside of refrigerators should be kept clean and uncluttered. Do not store tubes or bottles for long periods of time. Mold grows easily on labels, tube racks, and other containers left in the moist, cool environment of the refrigerator. Then, when the culturist removes reagents for use in cell culture, the contamination is spread to the working surfaces and the cultures.
3. Open water baths sometimes used in tissue culture laboratories to preheat medium are a major source of contamination. If these must be used, the water should contain a disinfectant and should be changed regularly.
4. The incubator is the optimal environment for the growth of many contaminants, as well as mammalian cells. The outside of plates can grow mold, as well as the water in the incubator or condensation on the doors, plates, seals, and surfaces. Water used for humidity inside incubators should contain a (nonlabile) germicide and surfaces should be wiped down with germicide frequently.
5. Trays that are removed from the incubator between experiments should be decon-

taminated or autoclaved and stored in a clean place until the next use.

6. Always clean up any spilled medium immediately since medium is the very best way to grow unwanted bugs, wigglies, balls of fluff, and any other not readily identifiable entities in the cultures.

7. Minimize movement of people in and out of the culture area. Needless to say (I hope), *NO* food or drink in the culture room.

WHAT CAN YOU DO TO GET RID OF CONTAMINATION IN CULTURES?

As much as people do not want to hear it, the best way to get rid of contamination in cultures is to (carefully) discard all contaminated cultures. This can sometimes mean all the cultures in the laboratory. In such unfortunate circumstances, all plates should be discarded and new vials thawed from the frozen cell banks or obtained from the supplier of the original cells. Obviously, if cell lines are frozen down frequently, as suggested above, this will be feasible most of the time. The only time that one should even try to "cure" an infection is if the cell line at stake is irreplaceable. This can happen if one is developing a cell line or selecting a mutant cell line or a line with significantly altered properties. There is always a window of exposure during this process when the contamination of a single plate can mean having to start over and the loss of months of work, or worse. In this case, an attempt at getting rid of the contamination may be worthwhile. The process of curing a cell culture of adventitious agents is not easy and the gain must be weighed against the risk of spreading the infection to other cell lines in the laboratory. It is expensive and time consuming. There is often a high relapse rate and one should be prepared for failure.

The first step is to reduce the burden of infectious agent as much as possible. Thus, one can physically remove a fungal particle, or extensively wash cultures contaminated with yeast, bacteria, or mycoplasma. Remember that the density of contaminant per milliliter of culture medium may be several logs higher than the cell density. After this washing step, one can treat the cultures with bactericidal or fungicidal or fungistatic agents. All of these are toxic to cells at some level. One hopes to kill the unwanted contaminant before permanently damaging the culture. With bacteristatic or fungistatic agents, the goal is to cause a large enough differential in growth rate to let the cultures outstrip the contamination. At the end of this treatment, the cells should be recloned by limiting dilution. In this way, even if the culture is still contaminated, as long as the contaminant is at a much lower density than the culture, it should be possible to find wells containing only cells. This cloning and subsequent passages should be done in the absence of any antibiotics or other antimicrobials. After two to three passages in antibiotic-free medium, the clones picked should be rescreened for the contaminant. This is particularly important if the contaminant was one that is difficult to detect visually, such as mycoplasma. For a mycoplasma contamination, the cell lines should be rescreened a month later and at frequent intervals for a year before it is declared cured.

For contaminating molds and yeast, some investigators have reported success by layering the cultures on a Percoll gradient, wherein the contaminants pellet at the bottom of the tube (Kruk and Auersberg, 1991; Overhauser *et al.,* 1990).

During this entire period, the cells should be handled in quarantine. It does not make sense to contaminate an entire laboratory in the process of trying to save one culture. The contaminated (or suspect) cells should be kept in a separate incubator and handled in a separate hood, if possible. If totally separate facilities are not available, the suspect cells should be handled at the end of the day and the hood, incubator trays, microscopes, and other surfaces should be cleaned thoroughly after use. The areas can then be allowed to dry overnight and the UV light left on in the hood to aid in decontamination.

REFERENCES

Hukku, B., Halton, D., Mally, M., and Peterson, W., 1984, Cell characterization by use of multiple genetic markers, *Adv. Exp. Med. Biol.* **172:**23–29.

Kotani, H., Phillips, D. M., and McGarrity, G. J., 1987, Malignant transformation of NIH-3T3 and CV-1 cells by a helical mycoplasma, *Spiroplasma mirum* strain SMCA *in vitro, Cell Dev. Biol.* **22:**756–762.

Kruk, P. A., and Auersperg, N., 1991, Percoll centrifugation eliminates mold contamination from cell cultures, *In Vitro Cell Dev. Biol.* **27A:**273–276.

Lubiniecki, A., 1990, Continuous cell substrate considerations, in: *Large-Scale Mammalian Cell Culture Technology* (A. Lubiniecki, ed.), Marcel Dekker, New York, pp. 495–513.

McGarrity, G. (ed.), 1977, *Mycoplasma Infections of Cell Cultures,* Plenum Press, New York.

Overhauser, J., Chakraborty, M. S., and Kelley-Card, L., 1990, Removal of yeast contamination from lymphoblast cultures, *Biotechniques* **8:**177.

Phillips, D. M. (ed.), 1977, *Electron Microscopy of Mycoplasma Infections of Cultured Cells,* Plenum Press, New York.

Phillips, D. M. (ed.), 1978, *SEM for Detection of Mycoplasma,* 2, in: Becker, R. P. and Joharic, O., *Scanning Electron Microscopy.* AMF: O'Hare, IL, pp. 785–790.

Van Diggelen, O. P., Phillips, D. M., and Shin, S., 1977a, Endogenous HPRT activity in a cryptic strain of mycoplasma and its effect on cellular resistance to selective media in infected cell lines, *Exp. Cell Res.* **106:**191–203.

Van Diggelen, O. P., Shin, S., and Phillips, D. M., 1977b, Reduction in cellular tumorigenicity after mycoplasma infection and elimination of mycoplasma from infected cultures by passage in nude mice, *Cancer Res.* **37:**3680–3687.

CHAPTER 8

Serum-Free Culture

The purpose of this chapter is to provide guidelines and suggestions for those investigators wishing to grow cells in hormone-supplemented, serum-free culture. It is neither exhaustive nor definitive. There are two major aims. The first aim is to call to mind specific conditions and viewpoints that may help orient the cell culturist to a slightly different perspective. With this aim in mind, points will be illustrated with results from work done on a specific cell line but which illustrate more general phenomena. The second aim is to cover technical details involved in setting up serum-free experiments, the preparation and handling of hormones, and so on that supplement the standard culture techniques used with traditional serum-containing cultures, which are covered in Chapter 5. Several elements of the role of serum in cell growth media and the substitutes used in serum-free culture are summarized. (See Table 4.3 for commercially available media specifically designed for serum-free culture.)

Finally, there is a large body of literature to which the investigator should refer in conjunction with this book, since it is impossible to cover all the details of growth of specialized cells in defined media. In some cases, the references have been selected to provide a different viewpoint or approach than that covered here. In addition, the references provided are designed to help provide more in-depth coverage of specific aspects of research using hormone-supplemented, serum-free medium. The four-volume series in *Methods for Cell and Molecular Biology* (Barnes *et al.,* 1984a–d) on serum-free cultures and the early papers on serum-free medium (Barnes and Sato, 1980a,b; Mather and Sato, 1979) might prove especially useful. The methods provided below were selected to provide a starting point to familiarize the investigator with the techniques and procedures discussed.

It should be emphasized at the outset that hormone-supplemented, serum-free culture is more than a method of saving money on the cost of fetal bovine serum. The very process of defining the hormone growth requirements for cells in culture provides valuable information on the control of growth, function, and differentiation of cells. These responses, although elicited under conditions that diverge markedly from the *in vivo* situation, provide valuable insights into the physiological control of these cell types. In addition, the ability to completely define and control the humoral environment of the cell allows for experimental design that simplifies interpretation of results from what is, at its simplest, still an extremely complex living system.

129

There are several reasons why serum would not be expected to provide an optimal environment for the growth and function of cells. First, no cell *in vivo* actually grows in serum but rather in a highly specialized local environment that may differ markedly from serum. Vascular endothelial cells grow in an environment most closely related to, but not identical to, serum. Not surprisingly, these cells tolerate much higher serum concentrations than many other cell types. However, for most cell types, undiluted serum is toxic and must be considerably diluted (five- to tenfold) to sufficiently decrease its toxicity. In addition, some factors in serum (e.g., high- and low-density lipoproteins, vitamin C) are unstable for freezing and prolonged storage. The absence or low concentrations of factors, such as hypothalamic releasing factors or heregulin, which may reach very high local tissue concentrations but may be represented in only minute amounts in the general circulation, makes it obvious that serum-supplemented media would have inadequate concentrations of hormones and growth factors to support the growth of some cell types. In addition to the lack of adequate levels of hormones, serum also may contain substances that inhibit growth, differentiation, and/or function. In fact, this has been shown to be true in several instances, in both primary cultures and established cell lines (Loo *et al.,* 1989). Serum may also stimulate the differentiation of cells to a nonmitotic state, making them impossible to maintain as an immortal cell line (Levi *et al.,* 1997).

Some of these roles of serum in culture and the serum-free substitutes available are outlined in Table 8.1. Historically, the advantage of being able to grow tissues in a completely defined medium was recognized at the turn of the century (Lewis and Lewis, 1911). The realization of this goal, however, awaited the development of complex media and the discovery, purification, and widespread availability of many of the growth and attachment factors required by cells in culture.

Attempts to grow cells in reduced serum have taken four major approaches. All have

Table 8.1
Comparing Serum and Serum-Free Growth of Cells

Serum	Serum-free
Provides hormones and growth factors necessary for cell function	Add hormones, growth factors, etc.
Provides attachment factors (e.g., fibronectin, vitronectin, etc.)	Coat plates with factor or polylysine
Acts as a pH buffer	Use organic buffers, e.g., Hepes
Binds and inactivates or sequesters toxic materials, e.g., organic compounds trace metals, cell-secreted products	Use highly purified H_2O and reagents; freshly prepared medium, hormones, etc.; add purified albumin
Contains binding proteins that stabilize and/or deliver hormones and nutrients to the cell	Add binding protein (a few are available or may be purified by published procedures), e.g., transferrin, ceruloplasmin
Provides nutrients	Use complex rather than simple growth medium
Contains protease inhibitors	Use trypsin inhibitor at subculture if trypsin is used for passaging cells; add compatible protease inhibitors to culture (e.g., aproteinin)
Contains differentiation factors	Add or omit, as desired, to regulate cell growth and differentiation
Contains factors that cause "senescence" in normal rodent cells	No crisis, cells continue growing
Supports the growth of many cell types, including fibroblasts	Can select for cell type of interest

been used in conjunction with continuing alteration and enrichment of the media compositions (Ham and McKeehan, 1979; Rothblat and Cristofalo, 1972).

131

**THE
SUBSTITUTION
OF DEFINED
COMPONENTS
FOR SERUM**

1. Adapting cells to serum-free medium with no hormone supplements (Evans *et al.,*
   1956).
2. Using serum fractions or sera depleted for certain classes of hormones (Nishikawa
   et al., 1975).
3. Supplementing serum-free medium with hormones, growth and attachment factors,
   and so forth (Barnes and Sato, 1980b; Bottenstein *et al.,* 1979).
4. Establishing and maintaining cell lines in serum-free media formulations (see
   Chapter 10) (Levi *et al.,* 1997; Li *et al.,* 1996b; Loo *et al.,* 1989; Roberts *et al.,*
   1990).

While these approaches may be useful to meet specific goals, this chapter deals initially with the third approach: supplementation of serum-free medium with defined factors
to replace serum for the growth of established cell lines. A section on reducing serum follows. Establishing cell lines in serum-free medium is discussed in Chapter 10, this volume.

THE SUBSTITUTION OF DEFINED COMPONENTS FOR SERUM

Monitoring cell growth provides a rapid and sensitive assay to assess hormone effects.
Growth is the sum of a complex series of events involving the regulation of many aspects
of cell metabolism. These include, in part, cell attachment and spreading, transport, macro-
molecular synthesis, mitosis and cytokinesis, cell movement, cell differentiation, and energy utilization. A small change in growth may thus reflect a major change in one specific aspect of cell function. This can be a disadvantage, in that changes in the growth rate of cells
in the presence of a hormone may be the sum of the regulation of multiple aspects of cell
function. Further elucidation of these various functions would then require other approaches. On the other hand, the use of growth as an assay for hormone response, because
of the breadth of the phenomena observed (indirectly), can frequently be used to identify
unexpected hormonal requirements and controls. It is also the logical endpoint to use if the
goal is to replace serum supplementation to support cell growth.

It should be kept in mind that the hormones that stimulate growth will not necessarily
be identical to those that stimulate functional responses or differentiation. In this light, when
testing for hormone effects on growth, both stimulatory and inhibitory effects should be
noted. There are several possibilities of how a hormone may affect these processes:

1. Promote growth and increase or decrease function.
2. Inhibit growth and increase or decrease function.
3. Have no effect on growth but increase or decrease function.
4. Inhibit growth by promoting differentiation to a nonmitotic phenotype (Li *et al.,*
   1996b).

If the desired endpoint is to optimize a specific cell function, for example, the secretion of a specific protein, then protein secretion should be measured and used to optimize
the medium supplements instead of cell number. Figure 8.1 illustrates the effect of various

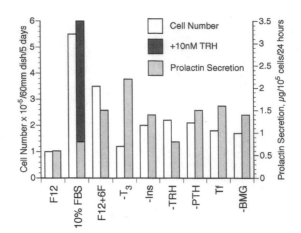

Figure 8.1. Differential hormonal regulation of growth and prolactin secretion by GH3 cells. One can easily see that prolactin secretion and growth are regulated independently. To optimize growth one might use serum or the six-factor (6F) supplement. But for optimal prolactin secretion, one would wish to eliminate T3 from the 6F condition and supplement with Thyrotropin Releasing Hormone (TRH).

hormones on the growth and prolactin secretion of GH3 cells. Figure 8.2 illustrates the effect of various hormone combinations on survival, steroid secretion, and human Chorionic Gonadtropin (hCG) binding in primary porcine Leydig cell cultures. Figure 8.3 shows growth inhibition of neuroepithelial precursor (NEP) cell line cells by Fibroblast Growth Factor (FGF) and Transforming Growth Factor Beta (TGFβ). While the growth inhibition

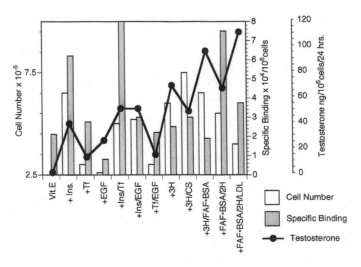

Figure 8.2. Differential hormonal regulation of survival and function in primary cultures of porcine Leydig cells. The conditions required to optimize (hCG) binding, cell number, and testosterone secretion are different. However, the maximal hCG binding seen in the insulin/transferrin (Ins/Tf) conditions might not be the preferred condition, since this represents an increase in binding over the levels in the freshly isolated cells, while the 3H (Ins/Tf and EGF) condition maintains the receptor level found initially.

133

**THE
SUBSTITUTION
OF DEFINED
COMPONENTS
FOR SERUM**

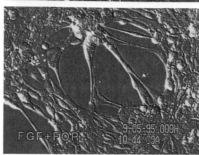

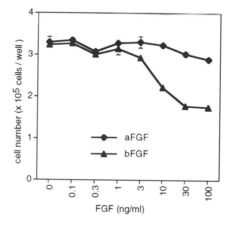

Figure 8.3. Differential hormonal regulation of growth and differentiation in the NEP cell line. The cells grow optimally in the conditions shown in the top photo. FGF inhibits cell growth (bottom graph), but, as seen in the middle photo, leads to a differentiation of the cells to a neuronal phenotype.

curves look similar, it is apparent from the photographs that TGFβ is killing the cells, while FGF is causing differentiation to a nonmitotic neuronal phenotype (Li *et al.,* 1996b).

These figures illustrate the above statement that the conditions for optimal growth are not necessarily those for optimal expression of function, and further, that various functions may be regulated differently. Understanding this, one can then design experiments to provide an environment that will maximize the expression of one function but potentially decrease other cell functions. Thus, "maximal" may not be "optimal" from the viewpoint of the cell, however the investigator looks at it.

Finally, the dose of hormone required for growth or function may vary from one cell line to another, from one function to another within the same cell type, or in different media or hormone supplements. This is illustrated in Fig. 8.4. TM3 cells show a 70-fold increase in growth in response to transferrin (TF), an iron transport protein, in Ham's F12 medium (with $FeSO_4$ omitted). Half-maximal response is seen at 25 ng/ml. These cells do

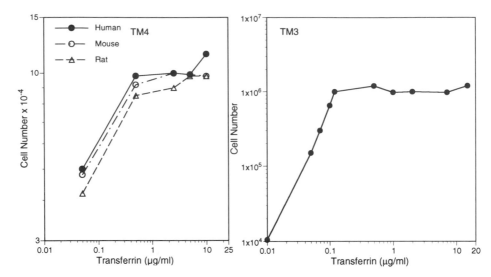

Figure 8.4. Growth response of several different cell lines to transferrin supplementation of defined medium. The TM4 cell line has a modest (at fivefold) growth response to transferrin, with no species specificity, while transferrin is a major growth promoter for the TM3 cell line.

not grow at low densities in this medium without TF but grow rapidly with just three factors, of which TF is the most critical. However, the Tr-1 cells, while sensitive (half maximal 50 ng/ml), do not have a stringent TF requirement, and grow slowly in 3H. Thus, only a fourfold increase in growth is seen with TF. The TM4 cells, in contrast, have a half-maximal response at 500 ng/ml TF. Thus, while three different cell lines may all respond to the same factor, the concentrations required and extent of growth stimulation may differ. In addition, nonhormonal components of the medium may affect cell response to a factor. Adding

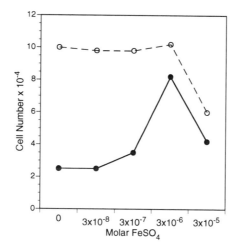

Figure 8.5. Iron and transferrin supplementation in defined medium. The addition of apotransferrin (○) allows the cells to grow at much lower iron concentrations, presumably by efficiently binding small amounts of iron in the medium and transporting it into the cells (●, without apotransferrin).

freshly prepared $FeSO_4$ to the medium will eliminate much of the growth response to TF by providing iron directly to the cell. Note, however, that free iron is more toxic to cells at higher concentrations than transferrin-bound iron (Fig. 8.5). There are also other interactions between nutrients and protein hormones (McKeehan *et al.*, 1981). This is one reason why it is important to select a complete medium before looking for growth factor responses in serum-free medium.

PREPARATION AND SELECTION OF MEDIUM

In serum-free culture the composition and preparation of the medium is far more critical than in serum-supplemented culture. This seems to be because of two major considerations. Many nutrients commonly provided by serum (i.e., vitamins) are not available in the simple media such as minimal essential medium (MEM) or Dulbecco's modified Eagel's (DME) medium. It is thus essential to use a more complex medium or a mixture of such media in serum-free culture work. Some of the more complete media available include Ham's nutrient mixtures (F10, F12, MCDB), RPMI, Waymouths medium, and so forth. (Table 4.1 lists the formulae for a number of commercially available media; the original source references are listed in the reference section at the end of this chapter.) In our experience, a 1:1 mixture of DME medium and Ham's F12 supplemented with bicarbonate, HEPES buffer, and antibiotics, if necessary, has proven to be excellent for serum-free work at both low- and high-cell densities. Figures 8.6 and 8.7 compare the growth of two cell lines in different media under hormone-supplemented, serum-supplemented, and serum-free conditions. A few simple experiments comparing several media or media mixtures for survival of cells in serum-free conditions can save a good deal of time and frustration in choosing hormone supplements later.

Second, serum protects against some toxic substances in media. It has been found that the water used to prepare medium is of critical importance in medium for use in serum-free culture work (see Chapter 4) (Mather *et al.*, 1986). Media should be prepared and stored in glass, Teflon, or tissue-culture-type plastics. All filtration and dispensing apparatus should use glass and silicon or tygon tubing (never latex). If media is filter-sterilized, the first 10–50 ml of medium may contain toxic components that washed out of the filter or appa-

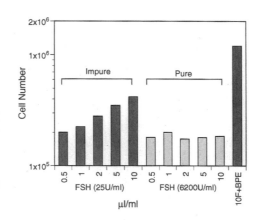

Figure 8.6. RL-65 growth response to impure and purified follicle-stimulating hormone (FSH) preparations. The RL-65 seemed to respond to FSH when an impure preparation was used. However, when highly purified FSH was tested, the RL-65 cells no longer responded, suggesting the active factor was a contaminant of the impure FSH preparation.

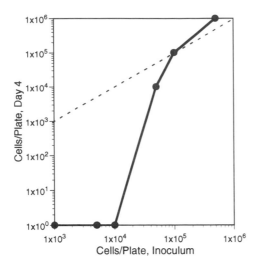

Figure 8.7. Plating inoculum density versus survival for M2R melanoma cells in serum-free and serum-containing medium. These cells, as with many, make autocrine factors that support their growth. If the cells are plated at too low a density, they will not survive and grow even in the presence of serum. As the medium supplements become more complete, the curve is shifted further to the left.

ratus and should be discarded. If glass flasks are to be used to prepare medium, it is best to purchase new flasks, wash them carefully with acid–ethanol or acetone, and rinse well with ethanol and distilled water. These flasks should then be reserved only for medium preparation and washed only with distilled water between each use. Washing flasks in a laboratory dishwasher can cause contamination by heavy metals from pipes, detergent not well rinsed off, or chemicals from other glassware washed at the same time. Any of these can lead to medium that is toxic to cells when grown without serum. Once the medium is prepared, it should be used within 2 weeks. Prolonged storage of serum-free medium results in poor cell growth or toxicity. If long storage time is necessary, some media (e.g., F12 or F12:DME, 1:1) can be frozen immediately after preparation (others will precipitate on freezing) and stored frozen for longer periods of time (1–2 months). Most standard commercially prepared liquid media have not been found adequate for serum-free work.

WATER

High-quality distilled water is a necessity for preparing media and solutions for serum-free culture. Double- or triple-glass-distilled water used soon (24 hr) after distillation is best. If possible, a distillation through acid permanganate is recommended to remove organic contaminants. This step is essential if deionized water is used on the still, as many deionizing columns seem to add a good deal of toxic organics to the water. There also are some column-based water purification units available commercially that work well for making serum-free media. These must be maintained according to the manufacturer's instructions to function efficiently. Water that is sufficiently pure for use in high-pressure liquid chromatography (HPLC) has been found to be good for preparation of media. All water should be transferred and stored using glass, Teflon, or tygon tubing and flasks.

The purification steps required will vary, depending on the water source, and even in the same area may vary throughout the year as the water source changes. Thus, it is safest to use the most extensive purification possible.

The sensitivity of various cell lines to water toxicity may also vary widely. Two cell

137

**PREPARING
AND TESTING
HORMONES
AND GROWTH
FACTORS**

lines, TM3 and TR-I, were carried in the laboratory in serum-containing medium with no change in growth or morphology. However, after 1 year of successful serum-free culture, difficulties arose with experiments using TR-I cells. Although the same double-glass-distilled water (water 2) had been used throughout, it was found that some changes had occurred that made medium prepared with this water toxic to TR-I but not other cells in serum-free medium. Preparing medium with this water and more highly purified water (water 1) immediately pinpointed the problem. As shown in Chapter 4, Fig. 4.2, the TM3 cells grew almost equally well in medium prepared from both water sources, but the TR-I cells could not survive in medium prepared with water 2.

It is obvious that this type of problem can be extremely difficult to pinpoint. It is thus best to prevent its occurrence by using the best quality water. The type of system purchased should reflect the type of input water available, the volume of use, and the cost. The ultra-pure filtration systems generally require less space and less maintenance but may be more expensive, since periodic replacement of the cartridges is necessary to maintain good quality water.

Other alternatives are possible and may be more convenient in your laboratory. We have made media from commercially bottled drinking water in France that was superior to the single distilled water available in the laboratory. However, even here the source made a difference and several brands were tested to find the best one. If possible, medium should be prepared using water from several alternative sources and compared on your cell line before a decision is made.

PREPARING AND TESTING HORMONES AND GROWTH FACTORS

The number of known hormones, growth factors, cytokines, and neurotropins (hereafter referred to as "hormones") described in the literature is increasing at a rapid rate. With the advent of genomics, many growth factors and receptors are cloned and expressed by homology or with no known biological response. It is thus possible to know a good deal about what cells produce or bind the factor before anything is known of the biology (Li *et al.,* 1996a). Defined culture systems should be invaluable in helping to describe the biological response to these new factors. Table 8.2 summarizes the solubilities and stabilities of some of the hormones, vitamins, and growth factors that are commercially available at this time. In many cases information on solubility, stability, and storage of these factors is provided by the supplier. There are also two excellent general reference works which are of great help in determining such parameters: (1) *The Merck Index,* and (2) *Handbook of Vitamins, Minerals and Hormones.*

STOCK PREPARATIONS

Since many of the hormones and growth factors must be solubilized and stored in different conditions at high concentrations, and all should be diluted to their final concentrations in media only immediately before use, the efficient preparation and handling of stock solutions is very important in making this type of culture less labor-intensive and frustrating. Stock solutions of hormones are prepared at 1000–5000× (for aqueous solutions) or 10–20,000× (for ETOH solutions) final concentration. These are then aliquoted into 0.5- to 1.0-ml aliquots and stored frozen or at 6°C, depending on stability, in glass or plastic

Table 8.2
Supplements that Can Be Used in Serum-Free Culture
and Active Concentration Ranges[a]

Defined supplement	Concentration
Tissue growth factors	
Epidermal growth factor (EGF)	0.1–10 ng/ml
Heregulin (HRG)	10–100 ng/ml
β-Cellulin	1–50 ng/ml
Acidic fibroblast growth factor (aFGF)	1–10 ng
Basic fibroblast growth factor (bFGF)	1–10 ng/ml
Multiplication stimulating activity (MSA)	1–50 ng/ml
Keratinocyte growth factor (KGF)	1–50 ng/ml
Platelet-derived growth factor (PDGF)	1–50 ng/ml
TGF family	
TGF-β1, 2, 3, 4, 5	0.1–10 ng/ml
Activins (A, B, C)	1–100 ng/ml
Inhibins (A, B)	1–100 ng/ml
Neurotropins	
NGF	1–10 ng/ml
GDNF	10–100 ng/ml
NT3	10–100 ng/ml
NT 4/5	10–100 ng/ml
SMDF	0.01–20 nM
BDNF	1–50 ng/ml
CTNF	1–50 ng/ml
Serotonin	0.05–0.2 μg/ml
Cytokines	
T-cell growth factor	0.01–1 μg/ml
Tumor necrosis factor	0.1–100 ng/ml
Lymphotoxin	0.01–1 μg/ml
Granulocyte-macrophage colony stimulating factor	0.01–1 μg/ml
Granulocyte colony stimulating factor	0.01–1 μg/ml
Erythropoietin (EPO)	0.01–1 μg/ml
Thrombopoietin (TPO)	0.01–1 μg/ml
Interleukins	1–100 ng/ml
IL-1α	
IL-1β	
IL-6	
IL-8	
IL-11	
IL-12	
Binding proteins	
Transferrin	1–5 μg
Ceruloplasmin	1–5 IU
BSA (highly purified)	1–25 μg
α$_2$-macroglobulin	0.1–5 mg
Follistatin	10–100 ng
IGF-1 binding proteins	0.01–10 μg/ml
Retinoid binding proteins	0.01–10 μg/ml
Hormones	
Insulin	0.1–10 μg/ml
Hydrocortisone	10^{-8} M
Testosterone	10^{-9}–10^{-7} M
Estradiol	10^{-9}–10^{-8} M
Progesterone	10^{-9}–10^{-7} M
Follicle stimulating hormone	1 ng–1 μg/ml
Leutenizing hormone	1 ng–1 μg/ml

139

PREPARING
AND TESTING
HORMONES
AND GROWTH
FACTORS

Table 8.2 (*Continued*)

Defined supplement	Concentration
Glucagon	10–100 ng/ml
Parathyroid hormone	10–100 ng
Leutenizing hormone releasing hormone	1–10 ng
Prostaglandin-E$_1$	10–100 ng
Prostaglandin-E$_{2a}$	10–100 ng
Somatostatin	10–500 ng
TSH	1–10 μg/ml
TRH	1–10 μg/ml
T3	10^{-11}–10^{-10} M
Calcitonin	0.01–10 μg/ml
Secretin	0.4–25 μg/liter
Caerulein	250–430 mg/liter
GLP	20–100 pg/ml
Gastrin	100–200 pg/ml
Substance P	0.1–20 μg/ml
Attachment factors	
Polylysine	Coat and wash
Fibronectin (Clg)	10 μg/ml, or coat
Collagen	Coat
Lamin	10 μg/ml or coat
Other defined additives	
Putrescine	100 ng–1 μg
Ethanolamine	1–10 μM
Phosphoethynolamine	1–10 μM
Selenious acid	2×10^{-8} M
Trace element mixture	See text
Thrombin	10–1000 ng
Aproteinin	10–100 μg/ml
Vitamins[b]	
Ascorbate (C) (unstable)	10–50 μg/ml
Retinoic acid (A) (unstable)	10–50 ng/ml
α-Tocopherol (E)	10–2000 ng/ml
Vitamin D	
Fatty acids and lipids	
Linoleic acid (unstable)	
Oleic acid	
Cholesterol	

[a]Concentration ranges given are those that usually have an effect when a highly purified, or pure, preparation of hormone is used in the absence of serum. The presence of serum, binding proteins, reducing agents, and the nutrient balance can shift some of the dose–response curves considerably. Impure compounds may be active at lower concentrations, but if the active concentration is much above the range given, the possibility of an active contaminant component of the preparation must be considered. Commercial sources for these growth factors are listed in Appendix 5. This is only a partial list and new factors are being discovered, cloned, and/or becoming commercially available each month.
[b]B vitamins are required for cell growth but are included in most nutrient mixtures. The vitamins listed are relatively unstable and therefore not included in most nutrient mixtures or may require supplementation by freshly made solutions if they are in the nutrient mixture.

polypropylene tubes. Where possible, solutions are made up in PBS. Medium should not be used to prepare hormone stocks or dilutions that are to be stored since some components of the medium will inactivate many factors. Frozen stock solutions should be aliquoted in amounts so that they are not frozen and thawed more than 3–5×. Working solutions are prepared at 100–1000× (aqueous) or 2,000–10,000× (ETOH) final concentration. These are to be used for a limited time, depending on the stability (e.g., insulin can be stored at 6°C for 3–6 weeks while retinoids are inactivated in solution within 24 hr when exposed to light and air). Within limits, working solutions can be mixtures of several compatible hormones [e.g., (EGF), FGF, and TF or testosterone and progesterone]. Many compounds that are only soluble in ETOH (or acid or base) at high concentrations are soluble in PBS in dilute concentrations. Thus, dilutions of progesterone ($< 10^{-4}$ m) can be done in PBS to minimize ETOH addition to the medium. Most cells will be damaged by ethanol concentrations of $>0.2\%$, so it is a good rule of thumb to limit total ethanol to this level.

Important: In most instances, one cannot add all hormones to the medium at final concentrations and store it. Medium with hormones stored overnight at 4°C and used the following day usually will not support cell growth.

Glass or plastic bottles used to store the medium will rapidly adsorb insulin, growth factors, and other added factors in the very-low-protein environment of most serum-free media. Media components, such as reducing agents, can also interact with and inactivate supplements. Some hormones may be quite stable in the storage conditions suggested, such as ethanol or low pH, but are not stable at neutral pH in medium. The degree to which storage will deplete the growth-promoting properties of the supplemented medium will depend on the supplements and the growth requirements of the cell. If storage of the medium is desired, the stored medium should be compared to the freshly supplemented medium for each cell type to be used. Since the inappropriate storage of supplemented media or hormone stocks is, in our experience, a common cause of variability and failed experiments, it does not seem worth the risk.

STERILIZATION

Care must be taken in sterilizing hormones to not lose all the hormone in the process. Those hormones that are ETOH-soluble can be made up in ETOH and placed in sterile glass or polypropylene vials. The ETOH itself will provide adequate sterility. Some factors are provided by the supplier in sterile form. These can be solubilized with sterile water or PBS in the vial they are packaged in and then aliquoted into sterile tubes. If filtration sterilization is necessary, this should be done at the highest possible concentration of hormone to minimize sticking to the filter. Millex disposable filters come sterilely packaged and have a reasonably small retention volume. A filter with low protein-binding capacity, such as the Millipore HA filters, should be used for proteins. Nonwettable filters are available and may minimize sticking. In some cases, the addition of BSA to the solution will minimize hormone loss, but BSA-only controls should then be used whenever these factors are tested. Where only very small amounts of hormones are available and one does not wish to add BSA, the hormone can be put into solution at high concentration, diluted in sterile solution ($>10,000$-fold dilution), and used directly in a medium supplemented with antibiotics. If care is taken to minimize growth of any minor level of contaminate in the stocks (i.e., by freezing), this is frequently adequate, but there is always some risk of contamination.

Care should be taken to use either glass or polypropylene (depending on the hormone; see Table 8.2) beakers, tubes, and pipettes for solubilizing, diluting, and aliquoting factors to minimize loss due to sticking. At best, one only knows what concentration of hormone one starts with, not what concentration is delivered to a culture, much less what remains after 1 hr, 1 day, or at the end of the experiment. This can only be ascertained by a direct measurement of hormone levels in the medium or cells. (This, however, is seldom done.)

A few words should be said about the source and quality of hormones. Except for synthesized or recombinant hormones, most are "hormone preparations." These will vary from one supplier to another and from one lot to another in activity and purity. It is possible that the growth-promoting activity in factor X is actually factor Y, present as a minor contaminant. If the purpose of an experiment is to study the effect of, for instance, follicle-stimulating hormone (FSH), the National Institutes of Health (NIH) FSH (or other FSH source) may be used; however, one will eventually need to show, as far as possible, that it is the FSH and not contaminating luteinizing hormone (LH), growth hormone (GH), thyrotropin (TSH), and so on that are eliciting the observed "FSH" effects. This can be done by using blocking antibodies, if possible, or comparing very highly purified hormone preparations from several sources, including recombinant.

Many suppliers will provide samples of several batches of a hormone for testing. One can then purchase a large amount of the lot that is the most effective and use the same lot in the series of experiments. If dose is critical, dose–response curves should be performed on each new lot of hormone. It is wise to maintain a skeptical attitude toward manufacturers' and suppliers' claims of activity and purity of hormones, particularly those purified from pituitary or sera that contain many active components. Manufacturers may also add buffers, BSA, or some other protein carrier to hormone preparations to act as a stabilizer. If so, you should have an excipient-only control. If you have the facilities, it is best to verify purity levels. Simple sodium dodecyl sulfate–polyacrylamide electrophoresis (SDS-PAGE) of the preparations can be an informative and sometimes a surprising and depressing experience. However, there is now usually a choice of suppliers for all of the commonly used attachment and growth factors. It is thus possible to choose the best source as to price, purity, and so forth.

Finally, it should be kept in mind that some hormones and growth factors (especially peptide hormones) show a species specificity for binding and/or activity. While it is frequently not possible to use hormones isolated from the homologous species, comparisons of the effect of hormones from the homologous species should be performed where possible. Heterologous hormones may be equally effective or effective at higher concentrations but much less expensive. Even in those cases where hormones' structures are identical, the active form may vary from species to species or from one tissue to another.

As an example of the points discussed above, Fig. 8.6 shows a growth-promoting effect of an impure FSH preparation on RL-65 cells that could be reproduced by pituitary extract but not by purified FSH. Obviously, in this case, FSH is not the true mitogen but rather another pituitary factor in the FSH preparation. Neutralizing antibodies or receptor antagonists, if available, can also be used to prove that the major constituent of a hormone preparation is truly the active factor. It is also desirable where possible to have a positive control. Thus, if the fact that cell "X" does not respond to FGF is important to the hypothesis, the FGF preparation should be shown to be active on a cell line known to respond.

SUBCULTURE AND SETTING UP EXPERIMENTS

Most cell lines are routinely carried in a serum-supplemented medium. Since these established cell lines have been grown in serum-supplemented medium for a period of years, it can be assumed that to a greater or lesser extent the serum supplement is providing the cells with the hormones required for growth and survival *in vivo* and/or the cells have adapted to the absence or reduced levels of some of the hormones required. The absence of sufficiently elevated levels of some hormones in 5–20% serum may in part account for the difficulty in establishing some cell types in culture. This point will be discussed in the following chapters on primary cultures and establishing cell lines.

Experiments on deriving serum-free medium to replace the requirement for serum are initially performed using stock plates that have been grown in serum-containing medium. This is to avoid selecting for cells with reduced hormone requirements. However, selecting for cells that do not require serum is a valid approach if the desired endpoint is to obtain a cell that requires a minimum of addition to the nutrient medium for growth. This approach will not help one understand what it is in serum that is required for the cells' growth nor will it aid in understanding the *in vivo* requirements of the cell.

There are several approaches to defining the hormone requirements for a given cell line. The method of choice will depend on the cell line. Several possibilities are outlined below. The initial step is to obtain conditions where the cells will survive and/or grow slowly for 3–6 days. In most cell types, this is partly a function of inoculum density. Figure 8.7 illustrates the survival of M_2R melanoma cells in serum-free medium as a function of inoculum density. For this type of cell, that is, one that will attach and survive in a serum-free medium with no supplementation, it is necessary to select only the proper inoculum density and begin testing hormones for growth-promoting effects. Once the optimal hormone supplement is found, the inoculum density required for survival will decrease. In some cases the plating efficiency of a cell line in hormone-supplemented, serum-free medium will be similar to that in serum, although this is not true for all cell types. This may be due to added requirements for attachment factors or growth factors needed only initially or at higher concentrations than those needed when cells are plated at high densities. Or it may be that autocrine factors produced by the cells and necessary for their survival or growth are unstable in the absence of serum or too diluted by the large medium-to-cell-volume ratio.

Many cells, both transformed and normal, are capable of producing substances that are required for their attachment or growth. However, some cell lines (e.g., GH3) will not survive even 24 hr or will not attach to the dish in serum-free medium. For these cells several initial approaches are used to obtain the minimal survival necessary for growth factor screening:

1. Use various purified attachment factors.
2. Precoat the dish with serum.
3. Plate cells in serum-containing medium for 12–24 hr; then remove this medium and wash the cells with serum-free medium before testing hormones.
4. Reduce serum concentrations to the point where the cells will survive but not grow.
5. Add test hormones, vitamins, growth factors, protease inhibitors, and so forth to this reduced level of serum.
6. After finding the mitogenic factors on the first-round supplement with these factors, reduce serum further and test again.

The various hormones can then be tested under these minimal conditions. When optimal conditions for growth are found, the serum (or preincubation step) can then be omitted and/or replaced with purified attachment and/or growth factors. A few general statements can be made regarding the requirements for optimal growth of cells in serum-free medium that will be of help in approaching the problem of defining serum-free media for a cell line:

1. To date, most if not all, cell lines that have been grown in serum-free media require insulin and transferrin. These two factors should be tested first and then included in all subsequent steps.
2. Most cell lines require one or more of the growth factors. These include EGF, FGF, Insulin-like Growth Factor (IGF), somatomedins, Nerve Growth Factor (NGF),

platelet-derived growth factor (PDGF), and so on (see Table 8.2). In addition, many new growth factors are being isolated each year which are specific for a certain cell type (Barnes *et al.,* 1991).

3. One can next test other classes of factors, including:
   a. Prostaglandins
   b. Steroids (first try each of the major classes of androgens, estrogens, and so forth, then try metabolites within these classes)
   c. Transport and binding proteins (e.g., ceruloplasmin, HDL, LDL, albumin)
   d. Pituitary hormones [or pituitary extract (Roberts *et al.,* 1990)]
   e. Thyroid hormones
   f. Fatty acids (complexed to albumin; liposomes) and complex lipids—HDL, LDL, and so forth
   g. Neurotropins or cytokines
   h. Protease inhibitors that are compatible with cell culture (e.g., aprotinin, α_2-macroglobulin)
   i. Fat-soluble vitamins such as vitamins A, D, and E

The hormone testing is best done in a stepwise fashion, testing new hormones in the presence of those found to be growth stimulatory. This is essential in some cases, as hormone effects are seldom simply additive. One cell line, shown in Fig. 8.8, shows little or

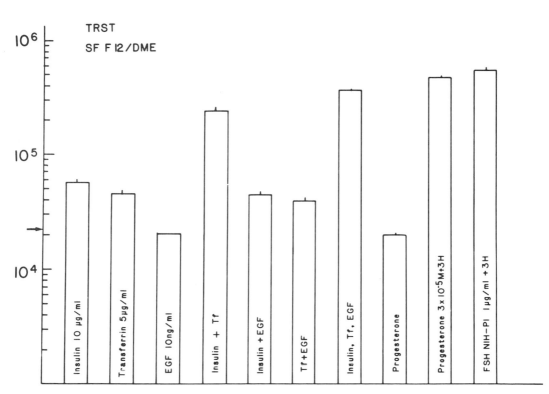

Figure 8.8. Addition of individual hormones and 3F. While addition of hormones such as EGF and progesterone individually has no effect, these factors will stimulate growth in the presence of insulin and transferrin (Tf). This type of response requires an iterative approach to optimizing media supplements.

no growth stimulation when insulin, TF, or EGF are added singly, but when all three factors are added together, the cells grow at a rate 50% of that seen in serum. In addition to these three factors, there are eight other factors that stimulate the growth of these cells in the presence of insulin, TF, and EGF, although none show any effect alone. When the cells are plated at high densities, they can grow without the addition of any factors.

Alternatively, some hormones and growth factors can stimulate growth by themselves, but their effects, when added together, cancel each other out or are inhibitory. An example of this is seen in the effect of retinoic acid and FSH on TM4 cells (Mather *et al.,* 1980). This cell line is derived from testicular Sertoli cells. Subsequent work in this and other laboratories has shown that Sertoli cells make transferrin, (IGF-1) and (TGF-α) (an EGF-like factor) *in vivo*. Thus, *in vitro* responses reflect the *in vivo* environment of these cells. In addition, some growth factors may replace the requirement for other less potent mitogens. This is the case with heregulin and FGF on Schwann cell growth (Fig. 8.9) (Li *et al.,* 1996c). In this case, one needs to determine whether receptors for both factors are present on the cell and whether both factors are produced in adjacent cells *in vivo* to determine whether both factors are likely to play a role in regulating the biology of the cell *in vivo*. It is important to remember, as stated previously, that a hormone may have a minor effect on growth but be of great importance in regulating some other function.

A complete replacement of serum by hormones ideally would allow for a doubling time and plating efficiency equal to (or in some cases greater than) that seen for that cell type in serum and the ability to carry the cell line through successive subcultures in the hormone-supplemented, serum-free medium. The dose of each hormone or factor should fall within the physiological range for that factor. It should be noted, however, that this is not always the case. In some cases a higher than physiological hormone level is required (e.g., insulin at 5–10 μg/ml), and in others a lower range (e.g., transferrin 0.5–5.0 μg/ml). Some cells do well for one or two passages in serum-free medium, but cannot be carried indefinitely (this suggests a hormone or nutrient is still missing but may be adequate for the experiments designed). There also is sometimes a longer lag phase or lower survival at plating in serum-free medium than serum-supplemented medium, but the subsequent growth rate is identical in the two media. Finally, one would want to test a more highly purified preparation of those factors originally tested in less pure form. For instance, test highly pu-

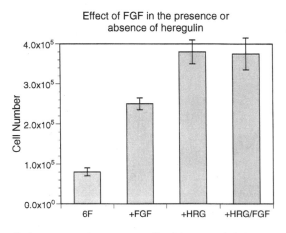

Figure 8.9. Schwann cells have a growth response to fibroblast growth factor (FGF) in the presence of six factors (6F), but there is no additional response in the presence of 6F+ heregulin (HRG). Understanding the biochemical basis of these types of responses can help shed light on the regulation of cellular function *in vivo*.

rified LH or hCG to see if it elicits the same response as higher doses of the NIH LH preparations. Additionally, the optimal dose of a given factor may vary in different media, for cells grown on different substrates, or in the presence of other hormones or growth factors.

It should be emphasized that a complete definition of the serum-free requirements may not be necessary for all applications. In some experimental systems a reduction of serum requirement by the addition of hormones or a submaximal growth rate may be adequate or even preferred for the studies to be undertaken.

REDUCING OR ELIMINATING SERUM

If the cell line is growing in the usual 5–10% serum-containing medium, one may wish to reduce the serum level. This makes the medium less expensive in many cases, simplifies purification by reducing serum proteins, and often improves product quality and/or cell growth. The general practice is to wean the cells by reducing the serum concentration over several passages (e.g., from 10% to 5, 2, 1, 0.5, 0.1%, to serum free), allowing two or more passages at each concentration if a reduction in growth rate is observed. This method often works best, but is time consuming. Eliminating or drastically reducing serum requirements for growth at high cell density over one to two passages is often a workable solution for cells that do not have very fastidious growth requirements.

The elimination or reduction of serum to below 1% usually necessitates the addition of hormones, growth factors, trace elements, and lipids (Ham and McKeehan, 1979). Because of the low concentrations of these components, the absence of binding and carrier proteins provided by serum, and the instability of the components in culture medium, these must be added to the media after filtration and just prior to use. The most commonly required additives are insulin (1–10 μg/ml), transferrin (1–100 μg/ml), and selenium (10–30 nM), however some cell lines have an added requirement for lipids. Many companies offer commercially available lipid supplements such as bovine lipoprotein (Pentex Ex-Cyte, Miles Laboratories; LipoMAX, Gibco/BRL #23000) or lipid-rich bovine serum albumin (AlbuMAX, Gibco/BRL #11020). Additionally, completely prepared supplement solutions containing insulin, transferrin, and selenium are available (GMS-S, GMS-A, GMS-G, and GMS-X Supplements, Gibco/BRL).

Some cell lines have more complex requirements for growth factors such as EGF, steroid hormones, or other small molecules, in which case the tissue from which the cell line originated can often provide clues to which components may be required for completely serum-free growth. A more complete description of deriving the appropriate hormone supplement in order to eliminate serum is given above.

For cell lines that are particularly fragile, or if the intended goal is suspension adaptation of the cell line, Pluronic F-68 (Mizrahi, 1975) or Pluronic F-127 have been found to be extremely effective in preventing shear-associated cell lysis. However, it should be noted that the addition of F-68 often prevents cells from attaching, especially in medium devoid of serum.

For some cells in serum-free medium, the addition of protease inhibitors to the medium will improve growth and/or survival. Protease inhibitors can prevent the degradation of endogenously produced or exogenously added growth factors by proteases secreted by the cells or released by dying cells in the culture. If one wishes to use protease inhibitors during the culture period, aproteinin or α_2-macroglobulin are less toxic than many of the small-molecule protease inhibitors.

Once the serum concentration has been reduced to the lowest possible concentration that supports adequate growth, a large population of the cells should be frozen down as a stock supply before further screening. Addition of 5–10% dimethyl-sulfoxide to the medium generally provides a satisfactory freezing medium (for fragile cells, one can also add 0.1% carboxymethylcellulose). All growth factors known to support growth and survival of the cells should be added to the serum-free medium in which the cells are to be frozen. It is preferable to freeze the cells in serum-free medium rather than reexpose the cells to serum. When the cells have been fully adapted to serum-free growth, or the hormone supplement worked out, freeze another 20–30 vials for future use.

CARRYING CELL LINES IN SERUM-FREE MEDIUM

A number of cell lines have now been established and are carried continuously in serum-free medium (Levi *et al.,* 1997; Li *et al.,* 1996b,c; Loo *et al.,* 1989; Roberts *et al.,* 1990). These lines cannot be grown in serum without inhibiting cell growth, changing the karyotype or altering the phenotype significantly. In the some cases, the cells die in serum-containing medium. Needless to say, these cell lines exhibit phenotypes not previously seen in serum-derived cell lines and represent new types of cells, opening up new avenues of research.

Carrying these lines may present a challenge to a laboratory not used to doing experiments in serum-free culture. All of the points described above for setting up serum-free cultures, and in the procedures below, apply. The cells must be subcultured on a strict schedule. Allowing overgrowth of the culture generally results in the loss of the culture due to a drop in pH, cell detachment, or depletion of a critical growth factor. Failure to neutralize the enzyme or wash out an enzyme inhibitor at subculture may also result in loss of the culture. The cells must be observed regularly and the required inoculum achieved at each passage. There must be an adequate supply of all required factors on hand, as certain cell types will die within a few hours of exhausting a particular hormone or growth factor. Frozen banks of cells representing various passages should be maintained.

While carrying cell lines in serum-free medium is more demanding, it can be done successfully and can provide important, new, and exciting options in choosing cell culture models for studying differentiation and hormone action. Most of the increased vigilance required for serum-free culture would also benefit cell lines carried using more forgiving standard cell culture techniques.

SETTING UP A SERUM-FREE GROWTH EXPERIMENT

The investigator should be familiar with the techniques described in Chapter 5 before undertaking these procedures. All the cautions and technical considerations described in Chapter 5 for setting up a growth experiment with serum apply here as well. In addition, further care must be taken in adding growth factors to the medium immediately before use, neutralizing and washing out the enzyme used to remove the cells from the plate, and keeping the medium at the appropriate pH and temperature to minimize cell damage. Time is of the essence when undertaking serum-free experiments. The goal is to have the cells out of the incubator for the minimum amount of time during the inoculation of the experiment. This is also true of hormones and growth and differentiation factors, some of which are quite labile at room temperature and reactive to light. Setting up several small experiments is often preferable to setting up one large experiment to address every conceivable question.

Materials

1. Cell cultures
2. Attachment factors. *Note:* If attachment factors are to be used, coat the plate with the purified attachment factor by adding the attachment factor in serum-free medium at the recommended concentration in a volume sufficient to cover the surface. Leave the plates in the incubator 1 hr to overnight and aspirate the attachment-factor-containing solution before use. If polylysine is used as the attachment factor, the plates must be washed well (2–4 times) to avoid toxicity due to unbound polylysine. (See below for cell-deposited matrix preparation.)
3. Trypsin or other dissociative enzyme
4. Enzyme inhibitor such as soybean trypsin inhibitor (STI) solution
5. Hormones or growth or differentiation factors prepared as 100–1000× stock solutions (see above). *Note:* These should be maintained on ice while in the tissue culture hood.
6. Basal medium of choice (see Chapter 4, Chapter 4, section on medium choice)
7. Tissue culture plates, precoated if necessary
8. Sterile disposable pipettes, pipettor tips, 15-ml conical tubes
9. Clinical or other low-speed centrifuge

Procedure

1. Label plates and place in the incubator. Remove excess fluid from plates preincubated with attachment factor.
2. Add serum-free nutrient medium to each plate.
3. Add water-soluble factors directly to the medium.
4. Add ethanol-soluble hormones by tilting the plate and dispensing the ethanol soluble factors into the deep portion of the medium. *Note:* This is especially critical when changing the medium on growing cells, since adding ethanol directly to a shallow layer of medium can result in cell death in the area of addition.
5. Put the plates containing the medium and all growth factor supplements in an incubator to equilibrate pH and temperature.
6. Enzymatically treat the stock plate of cells to be used.
7. Neutralize enzyme with STI or other suitable protease inhibitor.
8. Wash cells in medium to remove STI and any proteases not neutralized by the STI. *Note:* The STI contains lectins and other components that may prevent cell attachment or may be toxic. Washing out the STI is therefore important.
9. Dilute, count, and calculate the desired cell density for the inoculum. This should be minimal to avoid excessive dilution of the hormone supplements.
10. Remove the equilibrated plates from the incubator and inoculate cells.
11. Count cell inoculum at the beginning and end of the process. These numbers should be the same. If not, more care should be taken in steps 6–9.
12. Incubate the cells in the incubator for the desired time and then analyze the response.

Additional Notes on the Procedure

- Steps 1–5. When adding ethanol-soluble hormones, tilting the plate will provide rapid mixing and prevent all the added hormone from sticking to one spot (see Fig. 8.10). A control should be run, adding the maximum sum of ETOH added to the plates with the hormones. Most cells will tolerate 0.1–0.2% ETOH, with no effect on growth. Substances such as fibronectin should be added in a sufficient volume of medium to allow for an even coating of the dish. All hormones or a subset of hormones present in all conditions can be added directly to the medium before it is dispensed into the plates. If this is done, mix a sufficient volume of the medium and hormones in polypropylene tubes and dispense into the culture wells. If cells are to be added in a large volume, prepare the inoculum in serum-free medium and add hormone supplements at a level (e.g., 110% of final) to allow for dilution by the inoculum.

- Step 6. Placing the plates in the incubator while preparing the cells allows the pH and temperature to equilibrate and improves plating efficiency.

- Steps 7–9. In serum-free experiments, the trypsinization should be for the minimum required time to minimize cell damage. Low-temperature trypsinization (4°C) has been used by some investigators (Ham and McKeehan, 1979) to prevent uptake of trypsin by the cells. In this procedure the concentration of the trypsin activity is increased to compensate for the lowered temperature. Since there is no serum to provide proteins that neutralize trypsin, STI is added. The cells should then be washed well in medium (take up in 10–15 ml and centrifuge 1–2×) to remove the STI and any enzymes (e.g., collagenase, elastase) in the trypsin not neutralized by the STI and any contaminants (e.g., lectins) in the STI. *This washing step is important.*

- Steps 10–13. After washing, cells are diluted in medium to the required inoculum density. The use of an inoculum volume of 0.2–1.0 ml minimizes variability due to pipetting. However, if multiwell culture dishes are to be used, the inoculum volume may constitute such a large proportion of the final medium that the supplements should be added to the correctly calculated final concentration. The cells should be evenly dispersed over the dish with a swirling motion. This should be done after inoculating each two to four plates, as cells settle and adhere rapidly in serum-free medium. Uneven distribution of cells across the dish will increase the variability of counts on replicate plates. With good plating and counting technique, variability between replicate plates should be ± 5%. The inoculation of cells should be done as rapidly as possible (10–20 min) and the cells in the inoculum resuspended frequently to avoid settling or clumping. Prolonged time spent in

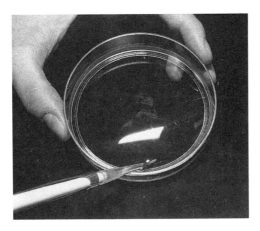

Figure 8.10. Ethanol solutions should be added to medium by tipping the plate and carefully pipetting the solution into the deepest part of the medium. This is especially important if the cells are already growing in the plate. Most cells will tolerate a final concentration of 0.1% ethanol.

inoculating and keeping the plates outside the incubator will result in a decreased cell survival in the first 12 hr. Experiments should thus be designed so that the number of plates is not too great to inoculate in this time period. The inoculum can be counted at the end of the plating to make sure there has not been excessive cell loss due to cells sticking to tubes or cell damage in the process of plating.

Preparing Plates Coated with Cell-Produced Matrix

1. Grow the cells that deposit the matrix (e.g., peritubular myoid, corneal epithelial cells) in the usual way for 1–3 weeks, with periodic medium changes, to allow a thick coating of matrix to be deposited.
2. Treat with sterile NH_2OH (0.01 N) until cells lyse and float off of the surface.
3. Wash with sterile PBS to neutralize (addition of phenol red to the PBS wash will allow visual confirmation of neutralization).
4. Store sterilely at 0°C in sterile PBS for up to 7 days, or at 37°C with medium 24–48 hr.

REFERENCES

Barnes, D., and Sato, G., 1980a, Methods for growth of cultured cells in serum-free medium, *Anal. Biochem.* **102**:255–270.

Barnes, D., and Sato, G., 1980b, Serum-free cell culture: A unifying approach, *Cell* **22**:649–655.

Barnes, D., Sirbasku, D., and Sato, G., 1984a, *Methods for Serum-Free Culture of Cells of the Endocrine System,* Vol. 2. Alan R. Liss, New York.

Barnes, D., Sirbasku, D., and Sato, G. (eds.), 1984b, *Methods for Serum-Free Culture of Epithelial and Fibroblastic Cells,* Vol. 3, Alan R. Liss, New York.

Barnes, D., Sirbasku, D., and Sato, G. (eds.), 1984c, *Methods for Serum-Free Culture of Neuronal and Lymphoid Cells,* Vol. 4, Alan R. Liss, New York.

Barnes, D., Sirbasku, D., and Sato, G. (eds.), 1984d, *Methods for the Preparation of Media, Supplements, and Substrata for Serum-Free Animal Cell Culture,* Vol. 1, Alan R. Liss, New York.

Barnes, D. W., Mather, J. P., and Sato, G. H., 1991, *Peptide Growth Factors, Part C, Methods in Enzymol.* Vol. 198, Academic Press, New York.

Bottenstein, J., Hayashi, I., Hutchings, S. H., Masui, H., Mather, J., McClure, D. B., Okasa, S., Rizzino, A., Sato, G., Serrero, G., Wolfe, R., and Wu, R. 1979, The growth of cells in serum free hormone supplemented media, *Methods Enzymol.* **58**:94–109.

Evans, V. S., Bryant, J. C., Fioramonti, M. C., McQuilkin, W. T., Sanford, K. K., and Earle, W. R., 1956, Studies of nutrient media for tissue cells *in vitro.* I. A protein-free chemically defined medium for cultivation of strain Leydig cells, *Cancer Res.* **16**:77–86.

Ham, R. G., McKeehan, W. L. (eds.), 1979, *Media and Growth Requirements,* Vol. 58, *Methods in Enzymol.* **58**:44–93.

Levi, A. D., Sonntag, V. K., Dickman, C., Mather, J., Li, R. H., Cordoba, S. C., Bichard, B., and Berens, M., 1997, The role of cultured Schwann cell grafts in the repair of gaps within the peripheral nervous system of primates, *Exp. Neurol.* **143**:25–36.

Lewis, M. R., and Lewis, W. H., 1911, The cultivation of tissues from chick embryos in solutions of NaCl, CaC_{12}, KCl and $NaHCO_3$, *Anat. Rec.* **5**:277–293.

Li, R., Chen, J., Hammonds, G., Phillips, H., Armanini, M., 1996a, Identification of Gas6 as a growth factor for human Schwann cells, *J. Neurosci.* **16**:2012–2019.

Li, R. H., Gao, W.-Q., and Mather, J. P., 1996b, Multiple factors control the proliferation and differentiation of rat early embryonic (day 9) neuroepithelial cells, *Endocrine* **5**:205–217.

Li, R. H., Sliwkowski, M. X., Lo, J., and Mather, J. P., 1996c, Establishment of Schwann cell lines from normal adult and embryonic rat dorsal root ganglia, *J. Neurosci. Methods* **67**:57–69.

Loo, D., Rawson, C., Helmrich, A., and Barnes, D., 1989, Serum-free mouse embryo cells: Growth responses *in vitro, J. Cell. Physiol.* **139**:484–491.

Mather, J., Kaczarowski, F., Gabler, R., and Wilkins, F., 1986, Effects of water purity and addition of common water contaminants on the growth of cells in serum-free media, *BioTechniques* **4:**56–63.

Mather, J. P., and Sato, G. H., 1979, The use of hormone-supplemented serum-free media in primary cultures, *Exp. Cell Res.* **124:**215–221.

Mather, J. P., 1980, Establishment and characterization of two distinct mouse testicular epithelial cell lines. *Biol. Reprod.* **23:**243–252.

McKeehan, W., McKeehan, K., and Ham, R., 1981, The relationship between defined low molecular weight substances and undefined serum-derived factors in the multiplication of untransformed fibroblasts, in: *The Growth Requirements of Vertebrate Cells in Vitro* (C. Waymouth, R. Ham, P. Chapple, eds.), Cambridge University Press, New York, pp. 223–243.

Mizrahi, A., 1975, Pluronic polyolsin human lymphocyte cell cultures, *J. Clin. Microbiol.* 2:11–13.

Nishikawa, K., Armelin, H. A., and Sato, G., 1975, Control of ovarian cell growth in culture by serum and pituitary factors, *Proc. Natl. Acad. Sci. USA* **72:**483–487.

Roberts, P. E., Phillips, D. M., and Mather, J. M., 1990, Properties of a novel epithelial cell from immature rat lung: Establishment and maintenance of the differentiated phenotype, *Am. J. Physiol. Lung Cell Mol. Physiol.* **3:**415–425.

Rothblat, G. H., and Cristofalo, V. J., 1972, *Growth, Nutrition and Metabolism of Cells in Culture,* Vol. 3, Academic Press, New York.

CHAPTER 9

Primary Cultures

While the majority of cell culture studies use established cell lines, there are some instances in which primary culture is preferred. The desire to study normal terminally differentiated cell types, such as neurons, myocytes, or T cells, *in vitro* obviously would require primary culture, since these cell types do not divide *in vivo*. In addition, it has proved difficult with some dividing cells to fully maintain their differentiated function *in vitro* through multiple passages. Quite often, this is not the result of "dedifferentiation," as is usually suggested, but rather is due to a deficiency in the culture conditions being used. In any case, there will always be a need to perform primary culture in order to study the properties of cells that are only recently removed from the *in vivo* situation in order to learn more about their functions *in vivo*.

The drawbacks of primary culture are that they are frequently time consuming to prepare and require the use of live animals or fresh tissue. There can be considerable variation from one preparation to another, particularly if prepared by different people. Finally, the cultures are continuously changing from the time the cells are removed from the body until they die or adjust to the culture conditions used. This can include changes in the mix of cell types in the culture, changes in cell shape, changes in cell–cell associations, and changes in the factors secreted from the cells and the receptors and other cell surface proteins present on the cells. These events must be taken into consideration in determining how long primary cultures can be studied and in interpreting the results obtained.

Primary culture refers to the cells that are placed in culture directly from the tissue of origin. These are called *primary cultures* until the first subculture. After the first subculture, they may be called *secondary cultures,* and thereafter, if continued passage is possible, a *cell line*. These cultures can contain mixed cell types or consist predominantly of a single cell type. However, given the complex interrelationships of cells in any organ or tissue, primary cultures seldom, if ever, consist exclusively of a single cell type. Attempts may be made to mechanically or enzymatically purify the cell type of interest during the tissue dissociation or to "clean" the cultures by culturing the cells in conditions that specifically inhibit or kill some major contaminant cell type(s). Alternatively, primary cultures can be maintained in conditions chosen to positively select for the survival of only one cell type. This approach is discussed further in Chapter 10.

There are undoubtedly more methods for putting cells in primary culture than there are cell types, and that is a great many indeed. In this chapter we will attempt to present general methods for some of the most commonly used types of primary culture and to refer to the volumes available on the isolation and culture of very specialized cell types. In general, the method chosen must take into consideration the desired balance between obtaining a high cell yield and minimizing damage to the cells to be cultured—between maximal yield and maximal purity. Considerations of the earliest time the cells are to be studied (after recovery) and the maximum time they are to be kept in primary culture might also determine the isolation method of choice. In preparing cells for primary culture, as in so many other aspects of life, you cannot have it all.

TUMORS

Tumors *in vivo* are by definition not long subject to the growth controls and regulatory mechanisms of their normal counterparts. When placed in monoculture, in the presence of serum, rodent tumor cells will often rapidly attach, spread, and divide. This may be due largely to autocrine factors produced by these cells that allow for rapid growth or to the activation of oncogenes that release the cells from normal growth control mechanisms. However, there are some rodent tumors, and many human tumors, that fail to grow easily *in vitro*.

The same biological laws are operative in either normal or neoplastic tissue: The *in vitro* microenvironment may be insufficient to sustain growth, either due to inappropriate factors contained in the medium, inappropriate substrate or matrix, inappropriate concentration of factors, or all the above. These factors need to be considered if there is difficulty growing a specific tumor *in vitro*. The special considerations for growing human cells (normal or tumorigenic) will be discussed.

Tumors are usually well vascularized *in vivo* and are often surrounded by fibrous stroma or connective tissue. Disaggregation of individual cells often requires both mechanical separation and enzymatic digestion. The digestion with certain enzymes (trypsin, crude collagenase) may prove toxic to some epithelial cells, although they are effective in removing and disassociating fibroblasts. In such cases a differential digestion with a crude collagenase containing STI, and then a gentler digestion with collagenase–dispase or thermolysin may prove more effective in obtaining viable tumor cells. Many tumor cells are less anchorage dependent than their normal counterpart and may be grown on plastic or substrates with reduced adhesive properties. Although tumor cells are considered mostly to be autonomous from normal growth control *in vivo*, once they are in monoculture they may depend on paracrine as well as autocrine factors for continued growth and survival and therefore need to be isolated and maintained at high density.

For example, we found this to be the case with a rare human esophageal carcinoma, which we initially isolated from a 1-mm biopsy specimen. We were able to isolate the tumor cells and obtain a cell line by maintaining them at high density in a small surface area initially (a single well of a 96-well microtiter plate) in 100 μl of medium containing insulin, transferrin, EGF, trace elements, and 0.5% serum. The low serum was sufficient to prevent initial fibroblast overgrowth, and in the presence of insulin, transferrin, and EGF the cells were capable of producing enough autocrine and paracrine growth factors to survive the initial transplant and divide. Conditioned medium from the parent cultures initially aided in passaging these cells.

For the most part, many tumors, particularly from rodent tissue, can be put into culture after careful dissection and mechanical mincing of the tumor with scissors. The B16 transplantable mouse melanoma is a good example. The method described below can be used for any relatively friable tumor type. Tumors with a great deal of connective tissue [such as the Engelbreth-Holm-Swarm (EHS) fibrosarcoma] will most likely require more extensive digestion to detach the cells from the matrix. This can be accomplished by increasing the concentration of enzymes, adding different enzymes such as elastase, and/or increasing the duration of treatment with enzymes. One can minimize damage to cells contained in such "tough" tumors by treating the tissue several times in succession, removing and washing the free cells after each treatment to minimize enzyme damage to the released cells, and combining the cells obtained at the end.

SETTING UP A PRIMARY CULTURE FROM A TUMOR

Materials for Dissection

1. B16 tumor about 5 mm in diameter (Jackson Labs, Cold Spring Harbor) (this comes packaged subcutaneously in a C57BL6 mouse)
2. Scissors, straight, 9 inch; curved, 9 inch; straight, 4-inch dissecting (iris); curved, 4-inch dissecting (iris)
3. Forceps, straight, 4 inch; curved, 4 inch
4. 70% ethanol in beaker (for soaking instruments)
5. F12–DME medium, 25–50 ml in sterile conical 50-ml tube with gentamycin (25 μg/ml)
6. 60-mm plastic tissue culture dishes, sterile
7. Pasteur pipettes, 9 inch, sterile
8. Rubber Pasteur pipette bulb, 2 ml (or pipettor)

Materials for Culture (all materials are sterile)

1. Tissue culture dishes: ten 60 mm or five 100 mm
2. F12–DME, with 25–50 μg/ml gentamycin added
3. 5 and 10 ml pipettes
4. Fetal bovine serum

Procedure for Dissection (this should be done outside the tissue culture room)

1. Kill the animal by CO^2 asphyxiation or by cervical dislocation.
2. Rinse the animal briefly in 70% ethanol.
3. Hold up the skin adjacent to the tumor and make an incision.
4. Carefully excise the tumor from the surrounding stroma and place in a 60-mm dish.
5. Remove any excess fatty tissue and necrotic tissue. Also try and remove all excessive fibrotic connective stroma. Place the tumor in a sterile 50-ml conical tube and transport it to a tissue culture hood.

Procedure for Tissue Culture

1. Place the tumor in a 60-mm dish and moisten with a few drops of medium with a Pasteur pipette.
2. Using at first a 9-inch pair of scissors and forceps (sterilize all instruments by soaking in a beaker of ethanol for 20 min), chop the tissue into small chunks.

3. With a straight iris scissors, cut the chunks into smaller pieces.

4. With a curved iris scissors, mince the tissue further until there are no discrete tissue pieces observable with the naked eye. It is important to keep the tissue moist with medium, but not so diluted that chopping the tissue becomes impractical.

5. At this point, the tissue pieces should pipette easily with a 5-ml pipette.

6. Add about 2–3 ml medium at a time to the tumor tissue and carefully pipette this suspension into a 50-ml conical tube.

7. Carefully resuspend the tissue pieces in a 50-ml conical tube, and QS with medium to 50 ml.

8. Spin down the tissue suspension by bringing the centrifuge up to speed (800 rpm) and immediately bringing it down again. This gets rid of excessive debris in the supernatant without allowing it to pack down in the tissue pellet.

9. Remove the supernatant and repeat steps 7 and 8.

10. At this point, if there is not a significant amount of fibrous tissue, the pellet can be re-suspended in 10 ml F12–DME and added to each of five 100-mm dishes (1 ml/dish) or ten 60-mm dishes (2 ml/dish). *Note:* If the tissue pieces are small enough to pass through the bore of a 1- or 2-ml pipette, then use this to pipette individual aliquots into each dish. If using a larger pipette, continue to aliquot only 1–2 ml/dish at a time, as drawing up 10 ml of this suspension and attempting to disperse it will result in a disproportionate number of tissue pieces in each dish, as the tissue tends to migrate to the miniscus end of the pipette.

11. Add F12–DME medium to a total of 5 ml in a 60-mm dish or 10 ml in a 100-mm dish, taking into account the addition of 5% serum. Add 5% fetal bovine serum.

12. Place in 37°C, 5% CO_2 humidified incubator.

13. After 24 hr, observe the plates under a phase contrast microscope.

14. If the cells appear to be attached to the plate and spreading out from the tissue, remove the medium and transfer it with any suspended tissue to a new dish. Add fresh medium with serum to the original plate.

15. If the medium appears quite yellow (acidic), wash the supernatant by centrifugation at 900 rpm for 3–4 min, remove the supernatant, and resuspend the pellet in fresh F12–DME with 5% serum in a new dish.

16. If there is much fibroblast growth, reduce the serum to 0.5% and add insulin (10 μg/ml) and transferrin (10 μg/ml).

17. When the plate has become 60–70% confluent, trypsinize as previously described and continue to maintain the culture at a 1:4 split ratio.

18. When the culture is expanded sufficiently, aliquots of the culture may be frozen for long-term storage. This should be done as soon as practical, usually at the fourth or fifth passage (or 20 population doublings) (see Chapter 5, freezing and thawing cells section).

The cells may be growing well by the fourth or fifth passage, but may be still a fairly heterogeneous population. It is at this point that the cells of choice may be selected by cloning (see Chapter 5, cloning section), by switching to a selective serum-free medium with defined hormones and growth factors, or by utilizing both methods. While switching to a selective serum-free medium that is supportive for the growth of melanoma cells (Mather and Sato, 1979) can inhibit growth of other cell types, it is possible that the primary cell type secretes soluble factors that support the growth of other cell types. This would lead to the persistence of a mixed cell population. It is therefore good practice to clone the culture whenever possible.

PRIMARY CULTURE OF NORMAL RODENT TISSUES

155

PRIMARY
CULTURE OF
NORMAL
RODENT
TISSUES

Isolating different cell types from the rat, mouse, or hamster, whether normal or tumorigenic, has the following advantages:

1. They are a readily available and relatively inexpensive source for tissue harvest.
2. Many inbred mouse strains are obtainable with well-characterized genetics. Increasing numbers of transgenic and gene deletion animals are also available for use in primary cultures.
3. There is a large body of literature using rodent-derived cultures.
4. Many cell types can be maintained in long-term culture without transfection or transformation.
5. The specific age, sex, and type of starting tissue are easy to replicate and control.

There are also disadvantages to using rodents:

1. They have a relatively unstable genome after freezing and thawing (Li *et al.,* 1996a).
2. They have a relatively small body size, limiting tissue availability.
3. They are not human.

In general, however, cultures of rodent cells have widespread utility and appeal, as they can provide valid and predictive models for their tissue of origin. Primary cultures from rodent tissues can routinely be obtained from adult, neonatal, and fetal animals. The advantage in using adult tissue is usually the volume of tissue available for harvest. Some experimental models specifically require adult tissue. But where the investigator is interested in establishing a cell line or simply long-term culture of a specific cell type, neonatal or fetal animal tissue has the advantage of providing a rapidly dividing population of cells whose long-term proliferative capacity is great and can frequently form immortal cell lines without terminal differentiation and/or senescence.

Two exercises are provided below that illustrate some of the approaches and problems in primary culture. For more detailed descriptions of primary culture of specialized cells, see the volumes edited by Wood (1992), Piper (1990), Freshney (1992), or Barnes *et al.* (1984a–c). The general procedure in trying to obtain a primary culture of any given cell type is outlined below. The preliminary experiments trying different factors, attachment proteins, and so forth can often be performed in 24- or 48-well tissue culture dishes to obtain the maximal data with minimal tissue. It is important to be able to see the cells in the dish and inspect them visually frequently.

1. Decide whether cell purity, cell yield, or minimization of cell damage should take priority in designing your protocol.
2. Determine what markers will be used to follow the cell type of interest and to determine cell purity and function throughout the culture. It is important to have more than one marker if the cells are to be studied in long-term primaries or a cell line is to be obtained, since individual markers may be maintained to different extents, depending on the culture environment.

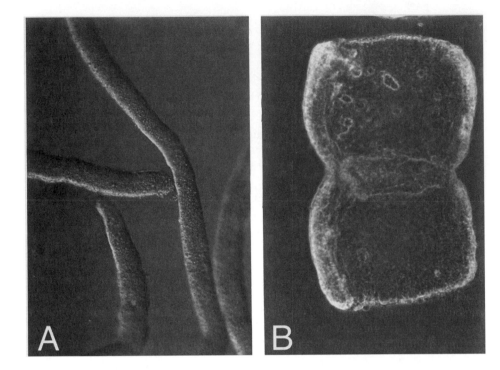

Figure 9.1. Dissecting tissues for primary culture. (A) Testicular tubules freed of interstitial tissue to be used for Sertoli cell cultures. (B) Dissected rat e9 embryonic neural plates to be used for the neuroepithelial cell cultures shown in Fig. 9.2. Finding and mechanically dissociating the tissue source for the desired cells in a primary culture is an important first step.

3. Decide on the animal species, sex, and age to be used. Locate the organ and cells you want to put in culture.

4. Dissociate the tissue containing the cell of interest (Fig. 9.1). Check for viability. If the cells of interest are not released from the tissue, try another dissociation method. If the cells have been released from the tissue but are dead, try a milder dissociation. If the tissue is still undigested, try different dissociation enzymes, more enzymes, and/or longer treatment times.

5. Inoculate plates at high and low density. Try several attachment factors. If live cells of the type desired are seen, continue incubating the cultures. Try several different media and different supplements, including serum and serum-free, hormone-supplemented medium (see Chapter 8) and conditioned medium (Fig. 9.2).

6. If the cells are dividing, try several different subculture regimens, for example, low and high density, frequent versus less often, trypsin versus collagenase–dispase. Keep watching the plates for several weeks; sometimes surprising things occur (for example, see protocol for primary culture of nonciliated cells, below).

7. Characterize the cultures. Determine the percentage of the cell type of interest in the culture and how this changes over the duration of the life of the cultures. If the cells can be subcultured or can be frozen and thawed for later use, then characterize the secondary or thawed cultures as well.

8. If none of the above has worked, try a different species (or strain) of animal, a different age of animal, or purposely coculture the cell type of interest with nearby cells or related (or not so related) feeder cell lines.

157

**PRIMARY
CULTURE OF
NORMAL
RODENT
TISSUES**

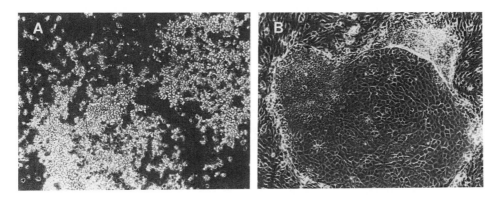

Figure 9.2. Primary culture of rat embryonic e9 neuroepithelial cells requires coculture or conditioned medium for survival. (A) All cells in the culture are dead 24 hr after plating in all supplements tried. (B) Culture supplemented with hormones and cocultured with an embryonic Schwann cell line. (The colony in the center is neuroepithelial and the surrounding cells, easily distinguished morphologically, are the ESC cells.)

9. If you have had success and the cultures are satisfactory for the need at hand, then publish. Publishing a complete description of the method used to derive a cell line is important for all those investigators who may later use the line. This initial publication should always be cited in any later publications using the line.

PRIMARY CULTURE OF 20-DAY-OLD RAT SERTOLI CELLS

In this protocol, the tissue dissociation conditions are chosen to minimize damage to Sertoli cell surface receptors while obtaining relatively pure preparations of testicular tubules through sequential chemical, enzymatic, and mechanical separation of the various cell types in the testis. The major contaminating cell type, the germ cell, can be removed easily from the cultures after the first day, if necessary. While yield is less important than purity in this protocol, a good yield is important, since the Sertoli cells are postmitotic at this age. The cells can be used for studies within 24 hr of plating and the cultures vary little over the first 1–2 weeks *in vitro* when maintained in serum-free, hormone-supplemented conditions. This procedure is described for use with day 20 rat tissue, but it has been used for rats 5–60 days old and for obtaining Sertoli cell cultures from mice, hamster, dog, pig, monkey, and human tissue with variations of the enzyme amount and time of digestion.

Materials

1. Five p20 Sprague-Dawley rats
2. 70% ethanol
3. Dumont #7 curved fine forceps
4. 4-inch iris scissors, straight and curved
5. Solution A: Glycine solution: 1 M glycine, 2 mM EDTA, 20 IU DNase (highly purified), and STI (0.002%) made up in Ca2+,Mg2+ -free PBS. Adjust the pH to 7.2 and filter sterilize.
6. *Note:* The 1 M glycine solution is hyperosmotic, and the EDTA is a chelator (e.g., binds calcium and magnesium). The DNase is not enzymatically active in the absence of di-

valent cations, but seems to bind to any released DNA and becomes active when the cells are returned to the medium. The STI protects against proteolysis. The alternation of hyperosmotic and normal osmotic solutions will lyse some cell types while leaving others viable. It also seems to frequently make it easier to subsequently remove cells from basement membranes during enzymatic dissociation. We have applied this procedure to good effect in isolating some types of cells from other tissues, but the results must be carefully monitored (Mather and Phillips, 1984).

7. Solution B: Collagenase–dispase solution: Collagenase–dispase (50 mg/ml, Boehringer-Mannheim) in medium with 20% (v/v), STI (1 mg/ml stock, Sigma); filter sterilize

8. 50-ml conical centrifuge tubes

9. F12–DME medium supplemented with insulin (5 μg/ml), transferrin (5 μg/ml), and EGF (5 ng/ml) immediately before use

10. F12–DME wash medium with 100 μg/ml gentamycin

11. Five 100-mm tissue culture dishes

12. Nitex nylon mesh 100-μm mesh size; cut into 6-inch squares and autoclave to sterilize

Procedure

1. Kill the animals by CO_2 asphyxiation on dry ice.

2. Soak the outside of the fur in ethanol and remove the testis.

3. Decapsulate the testis and remove as much of the testicular vein as possible.

4. Place the tubules in a 100-mm plate and add 15–20 ml of solution A.

5. Carefully disperse the tubules with the forceps while incubating for 10 min at room temperature. Gently pipette the tubule mass through a large-bore pipette (3-mm opening: use a 5-ml Pipetman pipettor with the tip cut off). The tubules will "unravel" to form a cloudy mass of tubules. Interstitial tissue is released (and severely damaged since interstitial cells collected at this point will not grow in culture).

6. Pipette tubules with a minimum amount of solution A into 50 ml F12–DME wash medium. Wash 3 ×, allowing tubules to settle by unit gravity. Discard wash.

7. Spin down tubules at 800 xg. Place tubules in a 100-mm dish with a minimal amount of medium. Chop to approximately 1- to 2-mm pieces with the curved scissors.

8. Add 15 ml of solution B and incubate at room temperature 10–20 min, until the peritubular myoid cell are seen to peel off the outside of the tubules and the tubules are in short pieces (1–2 mm long) and small cell clumps (20–30 cells). Place the digested tissue in a conical 50-ml tube, add wash medium, and let the clumps settle by unit gravity. Discard the single cells in the supernate.

9. Filter the tubular fragments through a 100-μM Nitex mesh cloth into a 50 ml tube.

10. Wash the fragments 3–5 × with wash medium.

11. Plate the cells in five 100-mm tissue culture dishes in medium supplemented with insulin, transferrin, and EGF. Cells should be attached and spread by 24 hr of culture.

PRIMARY CULTURE OF NONCILIATED LUNG EPITHELIAL CELLS

The following protocol provides an example of primary culture of a normal organ (rat lung). The tissue dissociation conditions are chosen to minimize cell damage and the culture conditions used to eventually select a single cell type, in this case a nonciliated bronchiolar epithelial cell (Clara cells), from the multiple cell types present in the primary lung explant (Fig. 9.3). Since the Clara cells are still dividing and differentiating at this age, one

159

**PRIMARY
CULTURE OF
NORMAL
RODENT
TISSUES**

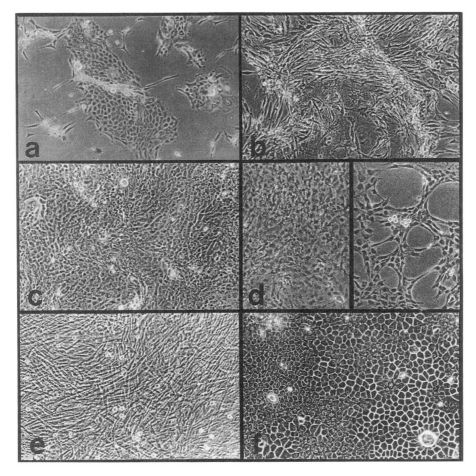

Figure 9.3. Primary cultures of newborn rat lung. (a) Defined medium selects for Clara cell survival. These cells can then be continuously cultured with the addition of (c) pituitary extract to form a cell line. Using other supplements such as (e) serum or (b, d) other mixtures of hormones, yields mixtures cultures of other cell types from the lung. (f) A 200× magnification of the established cell line.

can select for and expand these cells in the culture. While Clara cells do not become the predominant cell type in these cultures until the third week of culture, the cells can be passaged indefinitely and will form immortal cell lines of pure Clara cells (Roberts *et al.*, 1990, 1992).

Materials

1. Ten p.1-p4 Sprague-Dawley rats
2. 70% ethanol
3. Dumont #5 fine forceps
4. 4-inch iris scissors, straight and curved
5. Microdissection scissors (if dissecting out the bronchioles is desired)
6. Stereodissection microscope (if dissecting out the bronchioles is desired)
7. 50-ml conical centrifuge tubes

Table 9.1
Factors Used for Selecting Clara Cells from Primary Culture (11F)
and Minimal Factors Required to Grow the Established Cell Line (7F)

Factor	7F	11F
Insulin	1 μg/ml	10 μg/ml
Transferrin	10 μg/ml	10 μg/ml
Epidermal growth factor		5 ng/ml
Ethanolamine	1×10^{-4} M	1×10^{-6} M
Phosphoethanolamine	1×10^{-4} M	1×10^{-6} M
Selenium	2.5×10^{-8} M	2.5×10^{-8} M
Hydrocortisone	2.5×10^{-7} M	1×10^{-9} M
Forskolin	5 μM	1 μM
Progesterone		1×10^{-8} M
Triiodothyronine		5×10^{-12} M
Bovine lipoprotein (HDL)		0.5%
Bovine pituitary extract (15 mg/ml protein)	10 μl/ml	10 μl/ml

8. 11F isolation medium (see Table 9.1)
9. F12–DME wash medium with 100 μg/ml gentamycin
10. Ten 60-mm tissue culture dishes
11. Fibronectin (human, 1 mg/ml stock, Collaborative Research)
12. Collagenase–dispase (50 mg/ml, Boehringer-Mannheim) in medium with 20% (v/v) STI (1 mg/ml stock, Sigma)
13. DNase (1 mg/ml stock)
14. Bovine pituitary extract (commercially available, or prepared as in Appendix 3).

Procedure

1. Kill the animals by CO^2 asphyxiation on dry ice, or decapitate.
2. Rinse in ethanol.
3. Excise the lungs.
4. If dissecting out the bronchioles, the lungs should first be perfused by cannulating the trachea with a 25-g needle and perfusing with 1–2 ml of PBS. The periphery of the lobes can be cut into parallel 1-mm sections with a scalpel, and the bronchioles can then be dissected on a cold plate under a dissecting microscope.
5. If the bronchioles are not going to be dissected, remove the peripheral portion of the lobes to a 60-mm tissue culture dish, dissecting away the trachea and large bronchii.
6. Mince this tissue with the curved iris scissors until discrete chunks of tissue are not easily discernable (< 1 mm). The tissue should remain moist but not diluted with medium to facilitate mincing.
7. Add 3 ml of wash medium and 50 μl of collagenase–dispase solution per lung.
8. Place in a 37°C incubator for 50 min, removing every 10 min to resuspend with a 1-ml Pipetman and to observe the dissociation under a phase contrast microscope.
9. The fibronectin-coated dishes should be prepared at least 1 hr before and up to 24 hr prior to use. If they are not prepared prior to dissection, they may be coated during the enzyme incubation step.
10. After the incubation period, the cells will still appear clumped in the mass of tissue, but many cells will have been released. Add 50 μl of DNase solution to the dish, QS to 5

ml with wash medium, and add to a 50 ml conical tube. Repeat this step with each lung dissociated in this way.

11. Wash by centrifugation in a clinical centrifuge, 5 min at 900 rpm.
12. Resuspend the pellet in fresh wash medium and repeat step 11.
13. Repeat steps 11 and 12 until the supernatant no longer appears cloudy (3–4 times).
14. Resuspend the pellet in 10 ml 11F isolation medium. Constantly resuspend as 1 ml at a time is added to each of ten fibronectin-coated dishes.
15. Add 4 ml of 11F medium to each dish.
16. Place in a 37°C, 5% CO^2 humidified incubator.
17. After 24 hr, observe the cultures under a phase contrast microscope. If the cells are attaching and starting to spread, transfer the tissue in suspension to a new fibronectin-coated dish and add fresh 11F medium to the original plate.
18. Continue daily transferring the tissue in suspension, until it appears that cells are no longer dissociating, attaching, and spreading.
19. Observe the cultures daily and replenish spent medium. This is particularly critical in serum-free culture, as factors necessary for survival and growth can quickly become metabolized and their lack of availability can result in cell death within a few hours.
20. The medium should be changed every 3 days, even if the medium does not appear to be spent.
21. After 7–10 days, you will observe many cells dying off in the culture and the appearance of colonies of small cuboidal cells. After 3 weeks in culture, few if any of the cells in the heterogeneous population will remain.
22. Add 10 μl/ml bovine pituitary extract.
23. The cuboidal cells will begin to divide and grow exponentially within 24 hr and may then be subcultured with a 1:4 split every 4 days, in 15%, 0.22 μ-filtered, conditioned medium, taken from the parent dish.

——— SERUM VERSUS SERUM-FREE MEDIA ———

Both primary culture protocols outlined above use serum-free medium to select for and/or maintain the cell type of interest in the primary cultures. While many cells can survive in serum, we believe the serum-free approach has major advantages for primary culture as well as for growing established cell lines. Many advantages are the same as those outlined in Chapter 8. Serum is complex and undefined. In the tissue of origin, cells would come into contact with serum only during injury, so most cells respond to a serum-supplemented medium as they would to injury. Serum does supply growth factors and other constituents that will support the growth of a wide variety of cell types. However, many growth factors are required at higher concentrations than those in 10% serum (Roberts *et al.,* 1990). This may reflect the higher local concentrations reached in tissues when paracrine factors are produced locally and need travel only a short distance to reach their target cells. In addition, serum also contains growth inhibitors as well as differentiation factors, which may make it impossible to isolate and maintain a specific cell type (Loo *et al.,* 1989) in primary culture.

In the above protocol, we use serum-free, hormone-supplemented medium to isolate a single cell type from the lung by using medium that will support the survival of Clara cells only, and not support (or perhaps actively inhibit) other cells in the heterogeneous population of the initial explant (Fig. 9.3). By then adding a mitogen for the Clara cells, these cultures readily become cell lines. In the initial explant, the addition of serum stimulates fibroblast overgrowth. When serum is added to the established cell lines, previously unexposed to serum, the cells cease growing.

The Sertoli cells isolated in the Clara cell protocol will survive at high cell density without serum or hormones, but will retain their function for much longer with hormone supplementation (Rich *et al.,* 1983). Serum addition to the primary cultures leads to the proliferation of the small number of contaminating peritubular cells in the culture, which in turn secrete factors that modulate Sertoli cell function, changing the properties of the entire culture. Serum can also cause the loss of gonadotropin receptors in a lot-dependent fashion.

In any case, the use of serum is one more variable that is difficult to control, adding another level of variability to the innate variability of primary cell cultures. If possible, it is best avoided. Information on growing many different cell types in serum-free medium is given in the references at the end of this chapter and Chapter 8 (Barnes *et al.,* 1984a–c; Li *et al.,* 1996b; Mather and Phillips, 1984; Murakami *et al.,* 1985).

SPECIAL CONSIDERATIONS
FOR HUMAN TISSUES

Human biopsy material or material from resection or organ donors poses special problems. Usually, consent is required from the hospital review boards, attending physician, or surgeon and from the patient or the patient's relatives. In the case of biopsy, the first priority for tissue goes to the pathologist, often leaving little material for culture. The biopsy or resection often may occur at a time not convenient for the culturist, and delivery and preparation of the material must be planned well in advance with the surgeon. The risk of infection is another consideration when handling human material, and information regarding the screening for infectious agents should be obtained where possible. All human tissue should be handled with suitable precautions, including the use of a level 2 biocontainment hood and the autoclaving of all tissue culture waste.

Human tissue samples can vary considerably from donor to donor even if it is possible to obtain tissues from multiple donors of the same age and sex. The quality of the tissue also can differ depending on the time between removal and transport to the laboratory and the physical condition of the donor. It is frequently impossible to determine the entire course of treatment the patient has received before the tissue was removed. In the case of tumors, this treatment can be extensive and can affect the ability to obtain viable cells from the tissue and the phenotype of any cell lines obtained.

That said, some hints for culturing human cells are as follows:

1. Treat all tissue samples and resulting cultures as biohazardous material. Tissues and media should be treated with an agent such as bleach before disposal. Plasticware should be autoclaved before disposal. All work should be performed in a biosafety cabinet that recirculates and filters the air and vents it to the outside of the building.
2. Once culture conditions are established, freeze down cells early and often. Then results can be confirmed using cells from the same donor tissue. Human chromosomes are more stable to freezing than rodent chromosomes. Even cells such as human Schwann cells (Li *et al.,* 1996a), which proliferate rapidly initially, may abruptly slow down or cease dividing after a few passages. The number of passages that can be obtained from human tissue is dependent on the type of cell being cul-

tured, the age of the donor, and the culture conditions. The life span of almost all human cultures is limited by senescence, the as yet incompletely understood limitation on the life span of human cells *in vitro,* which applies even to widely used cell lines like WI-38.

3. Do as many experiments as possible using early passage or primary cells. If tissue is limiting, use smaller culture dishes and more sensitive assays to maximize the data that can be collected from each primary culture.

REFERENCES

Barnes, D., Sirbasku, D., and Sato, G. (eds.), 1984a, *Methods for Serum-Free Culture of Cells of the Endocrine System,* Vol. 2, Alan R. Liss, New York.

Barnes, D., Sirbasku, D., and Sato, G. (eds.), 1984b, *Methods for Serum-Free Culture of Epithelial and Fibroblastic Cells,* Vol. 3, Alan R. Liss, New York.

Barnes, D., Sirbasku, D., and Sato, G. (eds.), 1984c, *Methods for Serum-Free Culture of Neuronal and Lymphoid Cells,* Vol. 4, Alan R. Liss, New York.

Freshney, R. I., 1992, *Culture of Epithelial Cells,* Wiley-Liss, New York.

Li, R., Chen, J., Hammonds, G., Phillips, H., Armanini, M., Wood, P., Bunge, R., Godowski, P. J., Sliwkowski, M. X., and Mather, J. P., 1996a, Identification of Gas6 as a growth factor for human Schwann cells, *J. Neurosci.* **16:**2012–2019.

Li, R. H., Guo, W. Q., and Mather, J. P., 1996b, Multiple factors control the proliferation and differentiation of rat early embryonic (day 9) neuroepithelial cells, *Endocrine* **15:**205–217.

Loo, D., Rawson, C. Helmrich, A., and Barnes, D., 1989, Serum-free mouse embryo cells: Growth responses *in vitro, J. Cell. Physiol.* **139:**484–491.

Mather, J. P., and Phillips, D. M., 1984, *Primary Culture of Testicular Somatic Cells,* in: Barnes, D., Sirbasku, D., and Sato, G. (eds.), Cell culture methods for molecular and cell biology. Vol. II, Alan R. Liss, New York, pp. 29–45.

Mather, J. P., and Sato, G. H., 1979, The use of hormone-supplemented serum-free media in primary cultures, *Exp. Cell Res.* **124:**215–221.

Murakami, H., Yamane, I., Barnes, D., Mather, J., Hayashi, I., and Sato, G. (eds.), 1985, *Growth and Differentiation of Cells in Defined Environment,* Springer-Verlag, New York.

Piper, H. M. (ed.), 1990, *Cell Culture Techniques in Heart and Vessel Research,* Springer-Verlag, New York.

Rich, K. A., Bardin, C. W., Gunsalus, G. L., and Mather, J. P., 1983, Age dependent pattern of androgen binding protein secretion from rat Sertoli cells in primary culture, *Endocrinology* **113:**2284–2293.

Roberts, P. E., Phillips, D. M., and Mather, J. M., 1990, Properties of a novel epithelial cell from immature rat lung: Establishment and maintenance of the differentiated phenotype, *Am. J. Physiol. Lung Cell Mol. Physiol.* **3:**415–425.

Roberts, P., Chichester, C., Plopper, C., Lakritz, J., Phillips, D., and Mather, J., 1992, Characterization of an airway epithelial cell from neonatal rat, in: *Animal and Cell Technology: Basic and Applied Aspects* (H. Murakami, S. Shirahata, and H. Tachibana, eds.), Kluwer Academic Publishers, Boston, pp. 335–341.

Wood, J. N., 1992, *Neuronal Cell Lines: A Practical Approach,* Oxford University Press, New York.

CHAPTER 10

Establishing a Cell Line

Most investigators who use cell culture in their work will obtain their cell lines from other investigators or from a cell bank such as the American Type Culture Collection. It is certainly easier to obtain and use a previously established cell line than to create one. However, there may be instances in which no acceptable line exists with the desired properties, or is derived from the cell type or genetic background of interest. Transgenic and knockout animals also provide the opportunity to establish cell lines with known genetic anomalies. In these cases the investigators may wish to establish their own line in their laboratory. This chapter will deal with the techniques used to establish and characterize new cell lines from normal and transformed tissues.

TRANSFORMED CELL LINES

As previously mentioned, the best way to establish a cell line *in vitro* is to start with cells that are rapidly dividing *in vivo*. A tumor, by definition, is made up of cells that are rapidly dividing and will continue to do so, since they have escaped the normal growth control mechanisms of the cell. One can also purposely transform cells either *in vivo,* in order to obtain tumors that can be used to establish cell lines, or *in vitro*. Animal strains that have a high incidence of spontaneous tumors might be used to establish cell lines from these spontaneous tumors. Irradiation or chemical carcinogens can be used to induce tumors in animals. Alternatively, transformation may be induced *in vitro* by introducing transforming viruses or viral genes to primary cultures. A transforming gene with a regulable promoter (Hofmann *et al.,* 1992) may also be used to produce cell lines that can be switched from the transformed to normal phenotype.

More recently, transgenic animals have been produced that widely express transforming genes such as the SV40 T antigen. Tissues from these animals can be used to more easily establish cell lines *in vitro* (Noble *et al.,* 1995). Other transgenic animals have been created with targeted expression of transforming genes that predictably form tumors in specific tissues or cell types (Siegel *et al.,* 1994) or with gene deletions that predictably lead to tumor formation in specific tissues (Matzuk *et al.,* 1992). The main issues involved in obtaining cell lines from transformed cells is one of isolation of the cell type of interest and prevention of fibroblast overgrowth *in vitro* if these cultures are established in serum-containing medium. The approaches outlined above have led to the development of a number of cell lines derived from specialized cell types of exceptional interest. These lines are extremely valuable. It should be emphasized, however, that these cell lines are all, by definition, transformed and will therefore have properties significantly different from those of the normal tissue from which they are derived. These differences will of course frequently involve changes in the the growth regulation of these cells types.

TUMOR TISSUE

If a tumor is available and the goal is to establish a cell line from the tumor, the first step is to prepare a primary culture from the tumor tissue as described in Chapter 9 on primary culture. We prefer to use a serum-free medium, if possible, since this precludes any problems with fibroblast overgrowth and minimizes chromosomal instability and loss of some functional characteristics, as described in Chapter 8. If the cell type from which the tumor is derived has previously been grown in serum-free culture (e.g., primary culture or tumors from another species), then the previously used supplements make a good starting point. If the attempt to grow cells in serum-free medium fails, the addition of various amounts of serum may be tried starting with a small amount (e.g., 0.1%) and working up to the 10–20% level. This is best done in the presence of a "best guess" hormone supplement such as insulin and transferrin and trace elements. Several media may also be tried. Once the best conditions are determined for maximal survival in primary culture, the attempt to obtain a cell line can be made.

The main issues to keep in mind when trying to establish a cell line are:

1. Keep the cells in conditions that optimize their chance to grow (e.g., fresh medium, subconfluent cell density, periodic subculture, maximized medium nutrients and hormones).
2. Keep the cells at as high a density as is compatible with step 1 above.
3. Freeze cells down periodically during the process of trying to establish a cell line to insure against loss of the line owing to an accident and to allow for return to an earlier passage if desired properties are lost during the establishment of the cell line.

TRANSFORMING NORMAL CELLS *IN VITRO*

An alternative approach to obtaining transformed cell lines is to transform primary cultures of normal cells *in vitro*. This approach may be required when working with human tissues that do not grow indefinitely *in vitro* without some level of overt transformation. This approach also allows more control of the type of transforming agent and the extent of transformation than may be possible using cell lines derived from tumors. For example, one might conditionally transform cells using a temperature-sensitive SV40 T antigen or other controllable transforming agent.

To transform by viral or oncogene transfection of cells, one needs to choose the transforming gene to be used, the method of introducing the foreign DNA into the cells (if DNA is to be used), and the selection method to be used to select for the desired transformed phenotype. This has been covered extensively in other methods books (see Goeddel, 1991).

RODENT CELLS IN SERUM-FREE CULTURE

As discussed in Chapters 8 and 9, many rodent epithelial cell types will form cell lines and even immortal cell lines when the primary cultures are started, and the cells maintained, in serum-free medium. The critical point in obtaining normal cell lines in this fashion is obviously that that normal cell must be capable of dividing *in vivo*. One then tries to modify the *in vitro* environment in such a way that the *in vivo* conditions required for cell replication are duplicated. Clearly, any changes in the cell culture, such as the death of a large proportion of the cell type of interest during subculture, chromosomal changes in the cells due to freezing, mutagens, UV damage, or transfection with foreign DNA, would cast doubt on whether the cell line is subsequently completely "normal." Nonetheless, this approach had led to the establishment of a number of interesting functional, immortal cell lines from rodent epithelial cells, including FRTL thyroid cells (Ambesi-Impiombato *et al.,* 1980), SFME proastrocyte cells (Loo *et al.,* 1989), ASC and ESC Schwann cells (Li *et al.,* 1996), RL-65 lung Clara cells (Roberts *et al.,* 1990, 1992), and ROG pregranulosa cells (Li *et al.,* 1997) (Fig. 10.1). See Table 10.1 for these cell lines and the media and supplements used to grow them. Fibroblastic cell types seem to both require and do better in serum-containing medium, although serum-free media have been described for the growth of fibroblastic cell lines (Bettger and Ham, 1982). Mouse 3T3 (Todaro and Greene, 1963) cells, established in serum-containing medium from embryos, have been in culture for many years and are still used to study "normal fibroblasts."

These types of cultures are frequently somewhat more difficult to carry *in vitro* than standard serum-containing cell lines. They should be split on a strictly maintained schedule as to time between subcultures and density at subculture. Since these cell types may be absolutely dependent on a single growth factor for survival, the exhaustion of this factor in the medium or omission of this component may lead to death within hours. Many of these cell lines also require the use of attachment factors, another step in subculture. Additionally, the trypsin, or some other enzyme, used to remove the cells from the plate must be neutralized and washed out of the cells before replating. However, in spite of these difficulties, cell lines established in this manner have unique properties that lead to a new view of what regulates cell growth and differentiation *in vitro*.

HUMAN CELLS—LIMITED LIFE SPAN

To date, human cells in culture have not been seen to respond like the rodent cell lines described above. The equivalent human cell type, while frequently doing well in primary cultures in similar or identical media formulations to rodent lines, will still cease dividing after a finite number of population doublings. For example, normal adult rat Schwann cells can be selected for in a serum-free medium (Li *et al.,* 1996). These cells will form an immortal cell line without any overt transformation *in vitro* such as that described above. Human adult Schwann cells can also be selected for by using a similar serum-free medium (Fig. 10.2). This medium is also absolutely selective for the human Schwann cells, which grow well and can be carried through several passages. However, at this point, the growth slows and the cells are essentially nondividing. Currently, no normal immortal human cell lines have been established *in vitro*. All the available cell lines represent normal cells with limited replicative potential (e.g., WI-38 or HUVEC) or cell lines established from cells

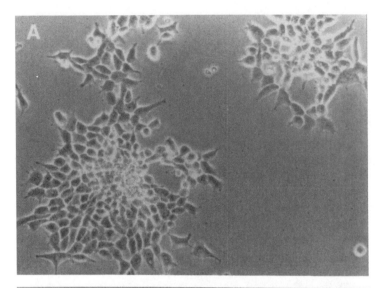

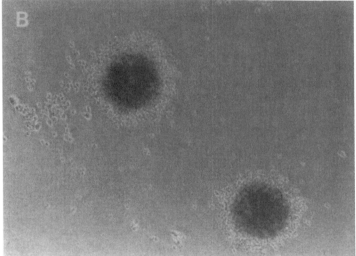

Figure 10.1. The ROG cell line, established from normal immature rat granulosa cells, maintains the ability to differentiate in response to FSH. The cells in (A) divide, flatten, and produce progesterone in response to FSH. After a 24-hr exposure to FSH, the cells are FSH dependent and will die apoptotically if FSH is withdrawn (B).

transformed *in vitro*, naturally occurring human tumors, or cells in a "hyperproliferative state" (Barnes *et al.*, 1981). The reasons for these differences between human and rodent cells are not well understood. It is possible that human and rodent cells are inherently different in their properties or that the culture conditions for human cells have not been adequately optimized to date.

Table 10.1
Rodent Cell Lines Established in Serum-Free or Low-Serum Medium

Cell line	Cell type	Medium[a]	Comment
B104 (Bottenstein)	Rat neuroblastoma	5 μg/ml insulin, 100 μg/ml TF, 20 nM Prog, 100 μM putrescine, 30 nM selenium, 10 μg/ml Clg	Basal media: BME
C6 (Huggins)	Rat glioma	2 μg/ml insulin, 5 μg/ml TF, 20 ng/ml FGF	Basal media: DMEM
eSC (Li)	Rat embryonic Schwann	5 μg/ml insulin, 10 μg/ml TF, 5 μg/ml Vit. E, 10^{-9}M Prog, 5 μM forskolin, 10 nM rhHRG3 μl/ml BPE	Basal media: F12–DMEM; requires laminin precoat
GH3 (Hayashi)	Rat pituitary carcinoma	5 μg/ml insulin, 5 μg/ml TF, 0.03 nM T3, 1 ng/ml FGF, 0.5 μf/ml PTH, 1ng/ml somatomdein C	Basal media: F12–DMEM
FRTL (Coon)	Rat thyroid epithelial	10 μg/ml insulin, 5 μg/ml TF, 10nM HC, 10 ng/ml GHL, 10 ng/ml somatostatin, 10 ng/ml thyrotropin	Basal media: F12
NEP (Li)	Rat neuroepithelial	10 μg/ml insulin, 10 μg/ml TF, 5 μM forskolin, 10 nM HRG, 10^{-9}M Prog, 5 μg/ml vit. E	Basal media: F12–DMEM; requires laminin precoat
M2R (Mather)	Mouse melanoma	5 μg/ml insulin, 1 μg/ml TF, 10^{-8} M testosterone, 10^{-9} M progesterone, 0.05 μg/ml FSH, 3 ng/ml NGF, 1 ng/ml LRF	Basal media: F12–DMEM
ROG (Li)	Rat ovarian granulosa	10 μg/ml, 5 μg/ml TF, 0.1 μg/ml Vit. E, 30 ng/ml rhActivin A	Basal media: F12–DMEM
RL-65 (Roberts)	Rat lung epithelial	1 μg/ml insulin, 10 μg/ml TF, 5 ng/ml EGF, 2.5×10^{-8} M aelenium, 10^{-6}M ethanolamine, 10^{-6} M phospho-ethanolamine, 5×10^{-7} M HC, 5 μM forskolin, 0.5% bLP, 8 μl/ml BPE	Basal media: F12–DMEM
SFME (Loo)	Mouse astroglial	10 μg/ml insulin, 25 μg/ml TF, 50 ng/ml EGF, 2.5×10 human HDL, 20 μg/ml	Basal media: F12–DMEM + fibronectin precoat
TEA3 A1 (Hayashi)	Rat thymic epithelial	10 μg/ml insulin, 10 μg/ml TF, 10 ng/ml EGF, 20 ng/ml cholera toxin, 10 nM dexamethasone	Basal media: WAJC 404A
TM4 (Mather)	Mouse Sertoli	10 μg/ml insulin, 5 μg/ml TF, 1 ng/ml EGF, 0.5 ng/ml FSH, 0.1 ng/ml GH, 10^{-8} M HC, 3×10^{-9} M Prog, 5 mU/ml ACTH, 50 ng/ml Vit. A, 0.0001% Vit. E.	Basal media: F12–DMEM

[a]TF, transferrin; Prog, progesterone; Clg, fibronectin; FGF, fibroblast growth factor; BPE, bovine pituitary extract; PTH, parathyroid hormone; GHL, glycine histadine lysine; FSH, follicle stimulating hormone; HRG, heregulin; HC, hydrocortisone; NGF, nerve growth factor; LRF, leuteinizing releasing factor; EGF, epidermal growth factor.

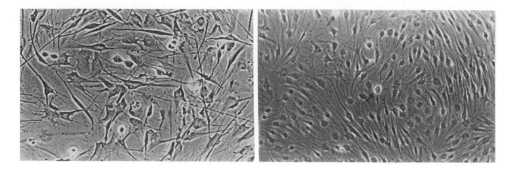

Figure 10.2. Phase photos of normal adult rat and human Schwann cell cultures in serum-free, hormone-supplemented medium. The rat cultures on the left are shown at passage 20 and continued dividing to establish a cell line. The human cultures are shown at passage 3. Their growth rate slowed appreciably at passage 5.

CRISIS AND SENESCENCE

In vivo, many cells that make up tissues in the organism have a finite life span. *In vitro,* it is possible to use selective culture conditions to extend the life of or in fact "immortalize" many cell types and still maintain the differentiated phenotype. Occasionally, cells may spontaneously transform in culture. When no attempt is made to transform the cell of interest, through special culture conditions or viral transfection, normal cell growth in serum-containing culture will begin to slow down by around the 30th passage and cease replicating altogether within the following few passages. This slowdown and eventual dying out of the culture is often referred to as "crisis" or "senescence": Few mitotic cells can be observed, although the culture may last several weeks and sometimes months in a nonreplicative state, as long as the culture conditions are maintained and the medium replenished. "Finite" cell cultures are generally best used between the 10th and 20th passages, and care should be taken to freeze down as many vials as practical between these passage numbers, or even earlier. Although freezing and thawing of cultures have been shown to sometimes cause genetic rearrangement, especially in murine cell cultures, the normal karyotype can best be preserved and phenotypic expression maintained when care is taken to freeze down cell cultures at these passages.

KARYOTYPING

Once a cell line has been established, it is wise to obtain a karyotype. This will confirm the species of origin, at least for those cells karyotyped, and determine the extent of gross chromosomal changes in the line. Karyotypes may vary from being near normal (i.e., the vast majority of cells in the culture have normal karyotypes) to aneuploid and extremely variable. It is not unusual to find some replication of a chromosome or piece of chromosome in some cells. While a normal karyotype is desirable, the presence of an abnormal karyotype does not preclude using the cells for *in vitro* studies, especially if it has been demonstrated that the cells are functionally stable. If a normal karyotype is required (e.g., cell lines to be used to create transgenic animals), then special care must be taken in handling the cells to minimize chromosomal changes.

The stability of the karyotype depends on the species from which the cell lines was derived, the growth conditions used, the way in which the cells are subcultured, and whether or not the cells are frozen. Loo *et al.* (1989) report the maintenance of near-normal karyotype in a mouse cell line, SFME, for over 100 passages *in vitro* in a defined serum-free medium. The introduction of serum into the culture caused chromosomal abnormalities to appear in the cells (Ernst *et al.*, 1991). We have found that rat Schwann cell lines had a stable karyotype when carried in serum-free medium for over 80 population doublings in continuous culture. However, freezing and thawing of the cells introduced some karyotypic changes in these cells (see Fig. 5.8). In contrast, we have shown near-normal and stable karyotypes in human adult Schwann cell cultures even after freezing and thawing, followed by further expansion of the population, although these cells have a limited replicative potential.

Karyotyping may be performed in the laboratory or the cells can be sent out to contract laboratories for karyotyping and analysis. A karyotype of a cell line established from normal adult rat Schwann cells (Li *et al.*, 1996) is shown in Fig. 5.8 (Chapter 5).

ESTABLISHING STERILITY

Since contamination is an ever-present threat in the early stages of establishing cell lines, it is assumed that any new line would have been monitored carefully for bacterial or fungal contamination during the process. It is especially important, however, to test the newly established cell line for mycoplasma before banking the line. Since these agents may influence many functional properties of the cell (see Chapter 7 for further discussion), interpretation of data obtained with the cell lines is dependent on their being shown to be free of such agents. Until such testing is performed, the cultures should be handled as if they were potentially contaminated. Contamination screening can be carried out as described in Chapter 7.

While there may be some rare instances in which a contaminated line may be used, it is very risky to decide to do so. There is risk of contamination of other cell lines being carried in the same laboratory or in other laboratories that may share a common tissue culture facility. A contaminated cell line should never be transferred to other collaborators, used in a common tissue culture facility, put through a FACS machine or other commonly used equipment, or carried into another laboratory, even temporarily, without specifically mentioning the fact that you are using a contaminated line to all others involved. Any paper published on work using the lines must specifically mention that it is a contaminated line and the identity of the contaminating agent.

CONFIRMING IDENTITY

It is important to check the identity of the newly established cell line. Indeed, it is wise to check the identity of any newly acquired cell line, or of any cell carried in the laboratory whose properties seem to have changed. The cell line should be checked for species of origin, tissue of origin, and the maintenance of specific properties. It is important not only to see that some cells in the population have the desired characteristics, but that most or all do.

Species of origin will of course have been confirmed in the cells that were karyotyped. However, it is wise to also use methods that can identify a minor contaminant of another cell type or of cells from another species. For species, these tests usually fall in one of two categories. The population can be tested by isozyme analysis. A selection of enzymes is chosen that have isozymes with different electrophoretic mobility patterns in the species most commonly used for cell culture. Extracts of cells are run on gels and the patterns compared between the cell line to be characterized and standard control lines or primary cultures from several species. Kits are available that provide all the necessary reagents and readout charts. Alternatively, one can use species-specific antibodies against cell surface proteins and use immunohistochemistry or fluorescence-tagged immunolabeling. This method has the advantage of being capable of screening large numbers of cells rapidly and of sensitively detecting low levels of contaminant. It is of course necessary to use both positive (of the expected species specificity) and negative control antibodies to maximize the chance of detecting contamination. Positive and negative control cell lines should also be run that will confirm the specificity of the antisera. With the advent of PCR, cell lines can be rapidly and periodically checked for expression of a specific mRNA or the presence of a gene. This can be adapted to detection of both microbiological contaminants and cross-cell-line contamination. In laboratories that carry a number of cell lines derived from the same parent cells and transfected with different genes, this may be the best means of detecting cross-cell-line contamination. It is possible to have cell line characterization performed by contract laboratories that will determine karyotype and species of origin. One such report is shown in Fig. 7.6 (Chapter 7).

All criteria used for determining the tissue and cell type of origin should also be rechecked at the time of banking or depositing the cell line (see following section). Again, it is highly desirable to include methods such as immunohistochemistry that will allow the determination of the presence or absence of the trait on a cell-by-cell basis.

It may seem like a lot of work to perform all these experiments to be able to say, "This line is what I think it is." However, the usefulness of all subsequent work on the line rests on the validity of the proper characterization of the cell line. Without this, improper conclusions, lost time, and embarrassing recall of papers may result. Here, an ounce of "prevention" is truly worth a "pound of cure."

"BANKING" THE LINE

All established cell lines should be cryopreserved or "banked" in order to assure a continuous supply of the cell line. It is wise to create a relatively large bank of the cells immediately after the above characterization of the cell line has been accomplished. Then, if any question arises in the future concerning the sterility or identity of the line, the investigator can return to the original bank. A bank of 30–50 vials is a good size. However, if the cell line is not immortal, or the investigators wish to use only early passage cells, then a bank of 100 or more vials may be advisable. Intermediate working banks of cells can also be laid down whenever the cells have been characterized for a particular function, recloned, or manipulated in any way that may significantly alter the properties of the line.

Cells to be banked should be grown up in several large plates or in a spinner. The cells should be removed from all plates and combined for all further manipulations. The goal is that each of the vials in the bank should contain the same number and distribution of cells from the total population. After the bank is prepared and stored in liquid nitrogen, a vial should be thawed and the plating efficiency of the thawed cells determined. One can thaw and wash the cells and do viability measurements and cell counts on the thawed cells. The cells can then be plated in the medium of choice and the cells counted after 24 hr to determine survival. The cells should be at least 90% viable directly after thaw, with greater than 75% of the cells surviving to the next day. If the banked cells do not meet these criteria, the conditions for freezing should be optimized and a new bank produced. Frozen cells will slowly lose viability over time even in the optimal storage conditions. Since the purpose of a cell bank is to have cells available over years, it is important to start with a cell bank of high viability or the cells may not be recoverable years later when the bank is needed.

As discussed in Chapter 5, the freezing medium, the rate of freezing and thawing, and the handling after thawing, including the plating density, all affect the viability. Newly established cell lines, particularly nontransformed lines, might be expected to be more fastidious about the conditions of freezing and thawing than cell lines that have been in use by many laboratories over many years and have experienced multiple freeze–thaw cycles. Therefore, it is wise to optimize freezing and thawing, using small aliquots of cells to determine the optimal conditions for the cell line of interest before putting down a large bank of cells.

It is often wise to store a portion of the banked cells in a separate facility, or at least in two separate storage tanks, to prevent loss in case of accident. Alternatively, one can place new cell lines on deposit with the American Type Culture Collection (ATCC) (Rockville, MD). They will store the cells and provide them to other scientists for a minimal fee. It is best if the cells submitted to the ATCC are extensively characterized as described above and the initial isolation of the cell line is published.

REFERENCES

Ambesi-Impiombato, F., Parks, L., and Coon, H., 1980, Culture of hormone dependent functional endothelial cells from rat thyroids, *Proc. Natl. Acad. Sci. USA* **77**:3455–3459.

Barnes, D., Van der Bosch, J., Masui, H., Miyazaki, K., and Sato, G., 1981, The culture of human tumor cells in serum-free medium, *Methods Enzymol.*

Bettger, W. J., and Ham, R. G., 1982, The critical role of lipids in supporting clonal growth of human diploid fibroblasts in defined medium, in: *Growth of Cells in Hormonally Defined Medium,* Vol. B (G. H. Sato, A. B. Pardee, D. A. Sirbasku, eds.), Cold Spring Harbor Press, Cold Spring Harbor, NY, pp. 61–63.

Ernst, T., Jackson, C., and Barnes, D., 1991, Karyotypic stability of serum-free mouse embryo (SFME) cells, *Cytotechnology* **5**:211–22.

Goeddel, D. V. (ed.), 1991, *Methods in Enzymology,* Vol. 185, Academic Press, San Diego.

Hofmann, M. C., Narisawa, S., Hess, R. A., and Millan, J. L., 1992, Immortalization of germ cells and somatic testicular cells using the SV40 large T antigen, *Exp. Cell Res.* **201**:417–435.

Levi, A. D., Sonntag, V. K., Dickman, C., Mather, J., Li, R. H., Cordoba, S. C., Bichard, B., and Berens, M., 1997, The role of cultured Schwann cell grafts in the repair of gaps within the peripheral nervous system of primates, *Exp. Neurol.* **143**:25–36.

Li, R. H., Sliwkowski, M. X., Lo, J., and Mather, J. P., 1996, Establishment of Schwann cell lines from normal adult and embryonic rat dorsal root ganglia, *J. Neurosci. Methods* **67**:57–69.

Li, R-H., Phillips, D. M., Moore, A., and Mather, J. P., 1997, Follicle-stimulating hormone induces terminal differentiation in a pre-differentiated rat granulosa cell line (ROG), *Endocrinology* **138**:2648–2657.

Loo, D., Rawson, C., Helmrich, A., and Barnes, D., 1989, Serum-free mouse embryo cells: Growth responses *in vitro, J. Cell. Physiol.* **139**:484–491.

Matzuk, M. M., Finegold, M. J., Su, J.-G. J., Hsueh, A. J. W., and Bradley, A., 1992, Inhibin is a tumor-suppressor gene with gonadal specificity in mice, *Nature* **360**:313–319.

Noble, M., Groves, A. K., Ataliotis, P., Ikram, Z., and Jat, P. S., 1995, The H-2KbtsA58 transgenic mouse: A new tool for the rapid generation of novel cell lines, *Transgenic Res.* **4**:215–225.

Roberts, P., Chichester, C., Plopper, C., Lakritz, J., Phillips, D., and Mather, J., 1992, Characterization of an airway epithelial cell from neonatal rat, in: *Animal and Cell Technology: Basic and Applied Aspects* (H. Murakami, S. Shirahata, and H. Tachibana, eds.), Kluwer Academic Publishers, Boston, pp. 335–341.

Roberts, P. E., Phillips, D. M., and Mather, J. M., 1990, Properties of a novel epithelial cell from immature rat lung: Establishment and maintenance of the differentiated phenotype, *Am. J. Physiol. Lung Cell Mol. Physiol.* **3:**415–425.

Siegel, P., Dankroft, D., Hardy, W., and Muller, W., 1994, Novel activating mutations in the neu proto-oncogene involved in induction of mammary tumors, *Mol. Cell. Biol.* **14:**7068–7077.

Todaro, G. J., and Greene, H., 1963, Quantitative studies of the growth of mouse embryo cells in culture and their development into established cell lines, *J. Cell Biol.* **17:**299–313.

CHAPTER 11

Special Growth Conditions

In this chapter, we will discuss some special considerations for growing cells in different formats from the more typical ones of monolayer plate cultures or small spinner cultures discussed in previous chapters. These will include miniaturizing conditions so that cell culture and growth assays can be adapted to a high-throughput format (Fig. 11.1), which uses a minimum amount of materials while generating quantities of data. The emphasis here is on efficiency, since the limit on how quickly an assay can be run is set by the biological doubling time of the cell. We also discuss scaling up cell culture systems for those applications where a large number of cells or large volumes of conditioned medium are needed for biochemical studies, protein purification, and so forth. These methods are still in the realm of what is desirable for use in a research laboratory. Industrial-scale culture is discussed in Chapter 12. Finally, we discuss growing cells in special configurations that change cell–cell associations (e.g., transwell dishes) or cell shape and cell matrix associations (e.g., in collagen matrix). It should be emphasized again that the increase in complexity of the systems used to grow cells does not alter the basic biology of the cells. All the conditions, caveats, and cautions discussed in earlier chapters still need to be taken into account when working with these more complex culture systems.

METHODS FOR HIGH-THROUGHPUT
——— ASSAYS FOR SECONDARY ENDPOINTS ———
CORRELATING WITH CELL NUMBER

Assays for cell growth work well as broad screens for factors that have a biological effect on specific cell types *in vitro*. An effect on growth may be the primary effect of the factor (e.g., a mitogen) or the increased or decreased growth seen may be secondary to altering nutrient transport, membrane stability, attachment, energy utilization, differentiation, progression throughout the cell cycle, or cell survival. The multiplicity of pathways involved in determining cell number at any given point in a culture increase the usefulness of

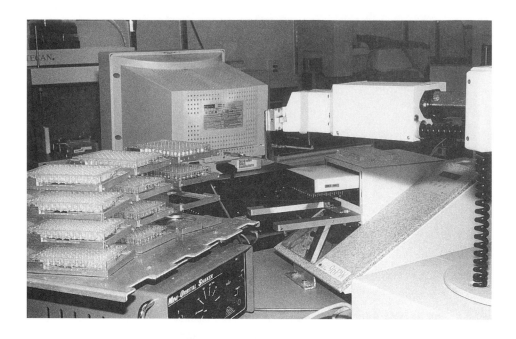

Figure 11.1. Robotics can be used for high-throughput assays. The photos show a robotics laboratory where high-throughput assays are run.

in vitro cell growth assays as a preliminary screen. In addition, there are a number of methods of assessing cell number that lend themselves to a high-throughput format.

The following methods lend themselves to a 96-well high-throughput assay format. The readouts vary in being more or less labor intensive, but the main part of the labor comes in setting up the cultures and adding the test substances. If these assays are to be done frequently, partial or complete automation of plate washing, addition of solutions, plate handling, and data collection can be explored as options. Some of these assays might be carried out almost entirely by robotics. The time required from setup to readout will be a function of the growth properties of the cell line used, the number of cells plated, the variability of the assay, and the level of sensitivity desired. In general, a 10% growth stimulation will not be apparent until two to three population doublings (DT) after the lag phase. This could vary from 24 hr after plating (for a cell with no lag and a 12-hr DT) to 5 days after plating (for a cell with a 2-day lag and a 36-hr DT).

The general range of sensitivity for each assay is given. The various ways in which test substances can give misleading readouts are discussed for each assay.

GROWTH OF CELLS IN 96-WELL PLATES

Materials

1. 96-well microtiter plates
2. Growth medium, 10 ml/microtiter plate
3. Multichannel (12 channels) pipette, 50–300 μl
4. Sterile pipette tips
5. Sterile reservoirs
6. Trypsin solution
7. Soybean trypsin inhibitor (STI) (1mg/ml; if serum free)

Procedures

1. Trypsinize the culture.
2. Resuspend the cells in 5 ml of growth media; count an aliquot.
3. If serum free, add STI (1 mg/ml) at 1:1 (v/v), resuspend in 5 ml medium, dilute to 10 ml medium in a 15-ml conical tube, and centrifuge for 3–4 min at 900 rpm. Aspirate supernatant and resuspend in 5 ml growth medium. Remove an aliquot for counting.
4. To the sterile reservoir, add enough growth medium, based on 6 ml/96-well plate in which only the center 60 wells will be seeded. When growing cells, particularly for longer times, this is necessary to maintain humidity for the inner wells and avoid artifactual readings due to evaporation from the outside wells. The outside rows of wells can be filled with sterile water or medium. One usually wants to run conditions at least in triplicate in 96-well plates since there is generally a higher percent variability in replicates in these plates than there is in the larger plates. There also can be considerable plate-to-plate variability, so a control should be run in each plate. Most reservoirs hold a practical maximum of 45 ml.
5. Having determined the cell number, add enough of the suspension so that 100 μl of the medium in the reservoir will contain 500–1000 cells. Prepare enough of the cell stock so that you have enough for one extra microtiter plate. This serves two purposes: (1) You have cells left to take a fairly accurate count after plating, and (2) there still will be a fairly even distribution and sufficient number of cells to seed the actual plates needed for the experiment.
6. Adjust the multichannel pipette to 100 μl.
7. Prewet the tips by drawing medium up into the tips and dispelling it back into the reservoir. Add 100 μl of sterile water or growth medium without cells to columns 1 and 12 and rows A and H.

8. Remove the 1st and 12th tip from the pipette.
9. Slowly draw the cell suspension into the tips and dispel it back into the reservoir. Pipette up and down this way for each row of 10 wells to be seeded. This will aid in keeping the cells in suspension evenly dispersed.
10. If a fixed concentration of the factor(s) is to be used, it can be prepared at twice the final concentration and added in 100 μl volumes to each well.
11. If a titration of the factor(s) is required, it should be diluted in growth medium and added to the individual wells in volumes not less than 1 μl or more than 20 μl.
12. Incubate the cells for 3–5 days (depending on growth rate, medium metabolism, and lag time) at 37°C.
13. Assay cell growth or death using one of the methods for measuring secondary endpoints discussed below.

CALCEIN-AM

Sensitivity: 10e4 - 10e6 cells per well

Calcein-AM (see supplier list in Appendix 5 for molecular probes) permeates the intact cell membrane and is cleaved by nonspecific esterases within the cytoplasm, hydrolyzing the acetoxymethyl ester, which becomes fluorescent. Substances that affect the production or activity of esterases will interfere with the assay.

Materials for Fluorescent-Based Calcein Assay

1. Calcein-AM (1 mg, dissolve in DMSO)
2. Fluorescent plate reader (Excitation: 485; Emission: 530).

Procedure

1. On the day of the assay, prepare the calcein-AM: Add 104 μl DMSO to the bottom of the tube of calcein (1 mg). Carefully resuspend the solution, by slowly pipetting three to four times.
2. To 10 ml of PBS in a reservoir, add 34 μl of the calcein-AM solution. Resuspend well with a 10-ml pipette. This is sufficient for one 96-well plate.
3. Remove the plate(s) from the incubator. If the cells are well adhered, shake the medium out of the microtiter plate over a sink or container and blot on a terry towel.
4. Submerge the plate in a container of PBS. Shake the plate and blot. Repeat once. A very gentle plate washer can also be used if the cells are firmly attached.
5. With a 12-channel pipettor, add 100 μl of calcein-AM solution to each well.
6. Incubate for 1 hr at 37°C.
7. Remove the plate, wash 2× by submerging in a container of PBS. Drain the plate and add 100 μl PBS to each well.
8. Read in a fluorescent plate reader (Excitation: 485nm; Emission: 530).

MTT REDUCTION

Sensitivity: $5 \times 10\ e3–10\ e6$ cells per well

MTT assays are based on the reduction of MTT [3-(4,5-dimethylthiazol-yl)-2-5-diphenyltetrazolium bromide], by live cells, to an insoluble (in aqueous solutions) dark purple formazan precipitate that can be read by a standard plate reader (Scudiero *et al.,* 1988).

1. As above for plating cells in 96-well plate
2. MTT (5 mg/ml stock; make fresh in PBS)
3. Orbital shaker
4. Standard plate reader (read at OD 560; reference, 690 nm)

Procedure

1. The cells can be trypsinized, plated, and incubated as in the general 96-cell method above.
2. To assay for cytotoxicity, grow cells to 50–100% confluency, then remove the medium, add 100 μl assay medium containing the drug of interest and incubate for an additional 24-48 hr.
4. Add 10 μl MTT/well, and incubate for 4 hr.
5. Shake the medium from the plate and blot (or carefully aspirate the medium).
6. Add 100 μl DMSO and shake on an orbital shaker for 5 min.
7. Read at OD 560, reference 690 nm.

OTHER DYE REDUCTION COLORIMETRIC METHODS

Other dyes such as MTS [3-(4,5-dimethylthiazol-2yl)-5-(3-carboxymethoxyphenyl)-2-(4-sulfophenyl)-2H-tetrazolium, inner salt] (Buttke *et al.,* 1993) and XTT [2,3-*bis* (2-methoxy-4-nitro-5-sulfophenyl)-5-[(phenylamino)carbonyl]-2H-tetrazolium hydroxide] (Roehm *et al.,* 1991) can be substituted for MTT. The MTS has a slightly better sensitivity than MTT or XTT.

[³H]THYMIDINE INCORPORATION ASSAY FOR DNA SYNTHESIS

Sensitivity: $500–10^6$ cells per well (depends on population doubling time and specific activity of the thymidine)

The measurement of thymidine incorporation into DNA during nuclear replication (or repair) can also be adapted to a high-throughput format by growing and labeling cells in a 96-well plate, directly harvesting the cells onto a glass fiber filter, which binds DNA. The incorporated [³H]thymidine is quantified using a scintillation counter or a gas-based counter. If this type of counter is available, thymidine incorporation can be adapted from the method described in Chapter 5 for use with 96-well plates and rapid counting methods so that it becomes a high-throughput assay. All the precautions described in Chapter 5 on handling and disposing of radioactive materials apply in this case as well.

The amount of incorporation of thymidine into DNA is proportional to the population doubling rate (in the absence of repair or multinucleation). Therefore, this is not a sensitive method for cells with very long division times, particularly when the total cell number per sample is limited by the size of the 96-well dish.

ALAMAR BLUE

Sensitivity: 1,000–80,000 cells per well

The alamar blue assay is based on detection of metabolic activity. The system incorporates an oxidation–reduction indicator that both fluoresces and changes color in response to a chemical reduction of medium in response to cell growth. Blue color is the oxidized form and red (also fluorescent) is the reduced. Since the dye is added directly to viable cells in the normal culture medium, continuous monitoring of cell viability over periods of several hours is possible. A real advantage of this procedure is that it requires minimal handling of the cultures and works with both attached and suspension cells.

Materials

Purchase working solution from manufacturer: Alamar Biosciences, 4110 N. Freeway Boulevard, Sacramento, CA.

Procedure

1. Grow cells of interest in 96-well plates as described above.
2. To assay at the end of the growth period, add 20 μl alamar blue solution (as per manufacturers instructions) per well and incubate plates as usual in the CO_2 incubator for 3–4 hr.
3. Read directly for absorbance at 570 nm or fluorescence at 560 nm excitation and 590 emission in a plate reader.
4. The plate can be reread at later points in the culture period as long as the cells remain viable (determine for each cell line; in our experience, 6–12 hr).
5. Linear range for absorbance: 5,000–40,000 cells per well. Linear range for fluorescence: 1,250–80,000 cells per well.

CRYSTAL VIOLET

Sensitivity: 500 (with solubilization)–1 million cells per well

This is a nonspecific protein stain. Agents causing an increase or decrease in protein per cell will give an increased or decreased readout in this assay without change in cell number. This is an inexpensive and easy-to-use method.

Materials

1. Crystal violet solution: Dissolve 5 g of crystal violet in 200 ml methanol and filter. Add 50 ml of formaldehyde (37%) and QS to 1 liter.
2. For low-density cultures: a solution of 0.1 M Na citrate (pH 4.2) and ethanol (50/50 v/v)

Procedure

1. Grow cells of interest in 96-well plates.
2. At the end of the experiment, remove the medium and add the crystal violet solution for 20 min at room temperature.
3. Wash by flooding the plate with water and flicking. Repeat three times.
4. Let dry.
5. If cells are confluent or at high density, the plates may be read dry in a plate reader at 540 nm.
6. If cells are low density, sensitivity can be increased by solubilizing the dye and reading the absorbency of the solution.

7. Solubilize the dye by adding the Na citrate/ethanol solution, 200 μl per well.
8. Seal the plate and shake until dissolved.
9. Read in plate reader at 540 nm.

ACID PHOSPHATASE

Sensitivity: 100–10,000 cells per well

This assay was originally set up for use with endothelial cells in 96-well plates (Connolly *et al.,* 1986). It is based on the use of a chromogenic substrate for acid phosphatase. Substances that regulate acid phosphatase levels or activity will interfere.

Materials

1. Reaction mixture: 0.1 M sodium acetate, pH 5.5; 0.1% Triton ×100; 10 mM *p*-nitrophenyl phosphate (an Abacus Cell Proliferation Kit (catalog No. K2020-1) with the necessary reagents is available through Clontech).

Procedure

1. Grow endothelial cells in 96-well plates.
2. Remove medium by aspiration.
3. Wash 2× with PBS.
4. Add 100 μl reaction mixture to each well.
5. Incubate 2 hr at 37°C.
6. Add 10 μl 1 N NaOH to each well.
7. Measure OD on plate reader at 405 nm.

— GROWTH CONFIGURATIONS FOR SCALING-UP — ATTACHMENT-DEPENDENT CELLS

Research laboratory or industrial production of intermediate-scale research materials may be required. If repeated production efforts will be needed using the same cell line, suspension growth in spinners is probably optimal. However, the investigator may not wish to heavily invest in equipment (spinner bases, spinners, fermenters, etc.) or take time to suspension-adapt a cell line or optimize the medium or cell line for a major production effort such as that described in Chapter 12. In these instances, one can produce large amounts of conditioned medium or cells with attachment-dependent cell lines using one of the other methods described later in this chapter.

SUSPENSION-ADAPTING CELLS

The purpose of suspension adaptation is to obtain cells that will grow as single cells unattached to a substrate. There are several approaches to reaching this goal: (1) alter the medium to eliminate or diminish the ability of cells to attach to the glass or plastic substrate; (2) select for cells that will not attach in the standard medium and substrate conditions; or (3) select for cells that will grow in suspension in the standard media when the surfaces available have been treated to prevent attachment (these cells may still attach to surfaces treated for tissue culture). The first approach has the disadvantage that the media devised to prevent cell attachment (generally with much reduced magnesium and calcium, e.g., Joclik's medium) are frequently suboptimal for supporting the secretion of high titers of desired proteins. The second approach is adequate for production cell lines but is more difficult than the third approach, and the resulting cell lines are less flexible. The third approach is generally (but by no means always) rapid and it results in a line that can still be grown in an attached fashion if desired for further manipulation, such as cloning or transfection.

The approach outlined below is designed to suspension adapt cells in this third sense, with as little alteration in other cell properties as possible. The one exception to this rule is that we sometimes have chosen to suspension adapt in a reduced serum (or serum free), hormone-supplemented medium in order to obtain a line that will grow continuously in these conditions. In at least one case, this strategy also improved our ability to suspension adapt the cells and to obtain a stable phenotype.

Medium: Standard serum-free F12–DME; supplement with:
- 2–10% fetal bovine serum (see Chapter 8 for a detailed discussion of serum reduction or elimination)
- Insulin, 5 μg/ml
- Pluronic F-68 (Gibco/BRL #24040-016), 0.1%
- HEPES buffer, 15 mM

Spinners: Use well-siliconized (see note) spinners. If the cells cake around the spinner shaft and on the sides of the spinner at the medium surface, the spinner is not properly siliconized. We prefer 250-ml spinners with 50- to 100-ml volume of medium. Spinner bases are set for a speed of 50–80 rpm. This should be just sufficient to keep cells in suspension and allow for good mixing of gases without undue mechanical damage to the cells. Set up two spinners in parallel. Carry both spinners, subculturing on different days, to prevent loss of the culture due to contamination. Suspension adaptation of most established (transformed) cell lines usually takes from 21 to 90 days. However, some normal cells will not suspension adapt without transformation.

Culture: Remove cells from the starting culture with trypsin and neutralize the trypsin with serum (or STI if using serum-free culture). Set up the spinners at $3–5 \times 10^5$ cells/ml. Check cells daily for growth and viability. More sodium bicarbonate solution may be added if the pH in the spinner drops below 6.8. After the first day or so, the caps on the spinner should be loosened to allow for increased oxygen and carbon dioxide exchange. On day 3 or 4, the cells should be counted and passaged. Initially, cells should be centrifuged and sufficient fresh medium added to the total cell pellet to bring the cell number to between 3 and 5×10^5/ml. As the cells start growing logarithmically to densities over 1×10^6/ml, they may be passaged by dilution of the suspended cells with fresh medium. If direct passage by dilution is to be used, the cell density should be such as to allow at least a 1:5 split. If cells cannot be split at this ratio or higher, the cells should be centrifuged, the old medium discarded, and sufficient cells seeded in fresh medium to obtain the desired inoculum density. Carryover of too much exhausted medium can progressively inhibit cell growth through the presence of toxic cell waste products or through nutrient depletion. If cells clump in the spinner, excessively large clumps should be allowed to settle so that they are not passaged.

After 2–10 weeks, the cells should be capable of logarithmic growth in suspen-

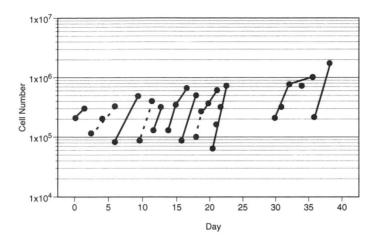

Figure 11.2. Suspension adapting a CHO cell line. Cells should be carried at high density until they start to grow vigorously. When cells can grow from 10^4 to over 10^6 cells/ml, they can be considered adapted.

sion to reach densities of $>10^6$ cells/ml when inoculated at densities of $5-10 \times 10^4$ cells/ml. Cell viability should remain at $>90\%$ throughout the growth period. At this point the cells are termed *suspension adapted,* even though they may still attach to tissue culture plastic, especially in medium containing serum or attachment factors. The cells may exhibit some clumping during prolonged growth in suspension, but this is not necessarily a disadvantage as long as the clumps are not so large as to cause necrosis of cells in the center due to nutrient limitation or to fall out of suspension.

The cell counts and passage times (arrows) are shown in Fig. 11.2 for the course of adapting a CHO recombinant clone. The cells were considered suspension adapted after 5 weeks.

SCALING-UP SUSPENSION-ADAPTED CELLS

Larger (1 to 3 liters, and greater quantities) spinners can easily be set up using the same cells and medium as used in the smaller (e.g., 250 ml) spinners (Fig. 11.3). If the cells have been suspension adapted, or grow easily in suspension (e.g., hybridoma cells), then obtaining increased volumes of harvest medium should be straightforward. Make sure all spinners are well sialated to prevent sticking of both the cells and the protein produced by the cells. Oxygen diffusion rapidly becomes limiting for many cells when grown in the larger spinners. If the cells are growing less rapidly or to lower density in large spinners than in the smaller spinners, decrease the volume of medium in the spinners to about one third the designated volume (e.g., 300 ml of medium in a 1-liter spinner). This increases the surface-to-volume ratio and allows improved oxygenation. Controlling pH may also be a problem, particularly if longer culture times are desired in order to allow accumulation of product in the medium. A bicarbonate solution can be added back to the culture as it becomes acid, to increase buffering capacity and keep the pH in the desired range of 7.0 to 7.5. It is also possible to purchase "instrumented spinners," which allow for aeration to increase oxygenation and for pH control. These are, however, more complicated, more labor intensive, and

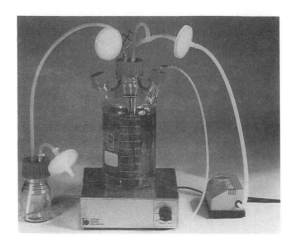

Figure 11.3. A spinner set up for growing large volumes of cells in suspension. This spinner allows for aeration through the pump at right. (Courtesy of E. Braun Biotech International)

more expensive than uninstrumented spinners. Spinner base controls should be set at around 40 rpm for 100-ml spinners and 60–80 rpm for the larger spinners. Sterile pluronic F-68 or some other polyol may be added at 0.1–0.5% (w/v) to the medium after filtration to improve cell resistance to shear. Set up a test run and determine the titer (weight protein/ml medium) of the desired product daily. Optimize the inoculum density and run time for optimal titer of the desired protein. Check to make sure the protein produced is intact and active.

ROLLER BOTTLES

Where large volumes of cells or culture fluid are required, and the cells of choice are not adapted for spinner culture or require attachment to a substrate for growth, roller bottles offer the investigator a convenient though labor-intensive method for "scaling up." A small roller rack can be placed in an incubator chamber, large incubator box, or warm room set to 37°C. Sterile, disposable roller bottles come in two sizes, with solid or vented (must be used in a CO^2 incubator chamber or with medium buffered for air) caps, and with surface area enlarged by "pleating" the surface, allowing for a higher cell-to-medium-volume ratio.

Cells should be plated in the roller bottles at a relatively high density (two- to fourfold more cells/cm^2 of surface area) (see Table 2.1) than that used in plates. This is to allow for the reduced initial growth rate experienced by many cells in roller bottles due to the decreased control over pH if the bottles are not grown in a CO_2 incubator. Plate in the same medium as used in tissue culture plates. It is best to have the medium prewarmed and equilibrated to the correct pH before putting it in the roller bottles. If the cells are sensitive to swings in pH, add an organic buffer and/or flush the airspace in the bottle with a 5% CO_2–95% air gas mix before sealing the cap. After placing the medium and cells in the roller bottle, seal the cap and place the roller bottle on the roller rack. Make sure the rack is level front to back, or uneven plating of cells will result. Set the rack at a slow speed (0.25 rpm) until the cells have attached and spread. Thereafter, for cells that attach firmly to the surface, a faster speed will allow more frequent exposure to fresh medium. A setting of 0.5 to 1 rpm works for most cells. If cells come off of the bottle in sheets at this speed, lower the speed.

If a known protein (e.g., an antibody or recombinant protein) is being produced, sample medium to assay protein levels daily. Medium should be harvested at the maximal titer or when the pH falls below 6.7. Medium can be harvested by pouring it out into a collecting vessel (in a sterile hood) and removing the remainder with a pipette. Be careful to wipe off any drops of medium on the outside to avoid contamination. Multiple harvests of medium can be obtained from the same roller bottle of cells by collecting the conditioned medium and adding fresh medium to the cells in the bottle. The second, third, and later harvests most probably will be after a shorter culture time than the first, since the cells will have had time to grow to a higher density.

If the production cell line is grown in serum-containing medium, purification of proteins from harvested medium can be aided by growing the cells to confluency in serum-containing medium and then using serum-free medium for the production rounds. This significantly reduces the protein in the harvest fluid. Make sure to check that the harvest medium contains the desired activity and that the cells remain viable throughout the production period. Supplementation of serum-free medium with insulin (5 µg/ml) frequently increases productivity and viability of cells in serum-free medium and seldom interferes with purification. See Chapter 8 for a more complete discussion of growing cells in serum-free medium.

Cells also can be harvested from roller bottles. Because of the large surface area involved, the fastest and cheapest way to harvest cells is by scraping. Special scrapers designed to get cells out of roller bottles are available. Some cells will be lost from the area around the neck of the bottle. Cells can also be removed enzymatically or with EDTA. In these cases, treat the bottle the way one would a plate. Return bottles to the roller rack during incubation with the trypsin or EDTA solution to be sure all cells are exposed to the fluid. Remove cells, wash the cells from the sides of the bottle with medium, and collect the released cells by centrifugation of the supernatant.

MICROCARRIER BEADS

If the laboratory is set up with the equipment necessary for spinner culture but the cell line to be used has not been (or cannot be) adapted to suspension culture, then microcarrier beads provide a good alternative to roller bottles. Microcarrier beads (e.g., Cytodex) can also be used in roller bottles to increase the surface-to-volume area the beads are growing on. The microcarrier beads stick to the roller bottle and the cells grow over both the bead and roller bottle surface. This technique can be used in laboratories that do not have equipment for spinner culture. However, harvesting of medium is still as labor intensive as in normal roller bottle culture. There are several types and manufacturers of microcarriers, including dextran beads, glass beads, metal springs, and collagen or agarose beads. Microcarriers commercially available at the laboratory scale are listed in Table 11.1.

The type of microcarrier selected will depend on the cell type, the way the medium or cells are to be used, and the container the bead–cell mixture is to be cultured in. Microcarrier cultures are most useful for harvesting secreted proteins from the medium, since the cells and medium are easily separated and the beads with attached cells can be used for further culture. Microcarrier culture is less convenient when cell harvest is desired, since the separation of the beads and cells is a laborious process and many cells will be lost. If cell harvest is desired, silica beads may prove easier to separate from the cells. If subcellular fractions are desired, direct lysis of the cells on the beads may be considered.

The most commonly used microcarriers are those made of dextran. They are relatively inexpensive, easily sterilized, and most cells stick readily to them. However, many secreted proteins will also bind to the bead material in the manner predicted for a dextran col-

Table 11.1
Comparison of Microcarrier Beads

Microcarrier (manufacturer)	Properties	Density (g/cc)	Size (μm)	Surface area (cm2/g)	Use/comment
Biosilon (Nunc)	Hydrophilic polystyrene	1.05	160–300	255	Nonporous, high growth, shipped sterile
Cytodex 1 (Pharmacia)	Cross-linked dextran	1.03	131–220	4400	General purpose, positive charge throughout matrix, good for production scale
Cytodex 2 (Pharmacia)	Cross-linked dextran	1.04	114–198	3300	Normal diploid cells, fibroblasts, good recovery of cell products
Cytodex 3 (Pharmacia)	Cross-linked dextran, denatured collagen layer	1.04	133–215	2700	Primary cultures, epithelial cells, high viability at harvest
Collagen (SoloHill)	Denatured collagen on copolymer plastic	1.02–1.05, 1.30	90–125, 125–212	475, 325	Good attachment, rapid growth, solid spheres
Glass (SoloHill)	Glass coating on copolymer	1.02–1.05, 1.30	90–125, 125–212	475, 325	Reusable, good for scale-up, high cell recovery
Plastic (SoloHill)	Polystyrene	1.02–1.05, 1.30	90–125, 125–212	—	Inexpensive, good cell viability, solid spheres

umn. The protocol below describes setting up a microcarrier culture, using these beads and CHO cells, for the harvest of conditioned medium.

GROWING CELLS ON MICROCARRIERS

Materials

1. Cytodex 3 microcarriers
2. PBS
3. F12–DME medium
4. Fetal calf serum (FCS)
5. Autoclave
6. Spinner
7. Spinner platform
8. CHO cells

Procedure

1. Rehydrate microcarrier as per instructions by adding 1 g of beads to 50–100 ml of PBS. Autoclave to sterilize.
2. Remove excess PBS and wash beads 2× with serum-free medium.
3. Transfer beads to a 500-ml spinner containing 250 ml F12–DMEM supplemented with 5% FCS (v:v).
4. Remove CHO cells from two nearly confluent 100-mm stock plates with trypsin. Treat with enzyme until a single-cell suspension is obtained.
5. Wash trypsinized cells in serum-containing medium and count.

6. Add 2.5×10 e5 cells to the spinner with the microcarrier beads.
7. Set spinner on spinner base and turn speed 20–60 rpm.
8. Turn up speed, loosen caps, and continue culture.
9. Check the beads daily for growth by removing a small aliquot and observing visually under the microscope. If the cell-to-bead ratio is correct, >90% of the beads should have at least one cell adhering to them and there should be few unattached cells remaining.
10. In order to quantitate cell growth, incubate a given aliquot of beads with alamar blue (see above, this chapter), or trypsinize cells from an aliquot of beads and count (see Chapter 5).
11. To harvest the conditioned medium, turn off the spinner and allow the beads to settle by unit gravity. Pour out the medium through 50 μm sterile Nytex cloth to catch any floating beads or remove them by centrifugation. To continue growing the cells, add fresh medium to the beads immediately. Do not allow the beads to settle for any longer than necessary (a few minutes) or the cells may die from oxygen deprivation.
12. Conditioned medium can be harvested at set intervals (2–6 days) and fresh medium added to prolong cell viability so that repeated harvests may be obtained from the same cells. This helps amortize the initial cost of the beads. If repeated harvests are desired, the harvest interval should be the longest possible without exhausting the medium (i.e., harvest before the pH falls below 6.8). Cells that exhibit contact-inhibited growth control will reach resting phase when confluent on the microcarrier beads.

GROWING CELLS IN HOLLOW FIBERS

Hollow fiber systems are designed to provide a very high cell-number-to-medium-volume ratio. The cells reside inside the apparatus and cannot be observed visually or sampled easily. Hollow fiber production is appropriate only where culture fluid is desired, since it is difficult to quantitatively recover the cells from these systems. These systems are designed primarily for the continuous production of harvest fluid. They make the most sense where ongoing production of a reagent (e.g., a monoclonal antibody) is desired. The purification should then be carried out on a regular basis, or even in-line with the harvest, so that one does not accumulate large quantities of harvest fluid, as with the roller bottle or spinner harvest. This method also works well for proteins that are degraded or unstable in cell culture medium at 37°C, since the medium can be harvested on an ongoing basis, thus minimizing the amount of time the protein is exposed to the culture milieu.

Hollow fiber systems come in a variety of sizes from those designed for large-scale production to small systems that may be appropriate for research into the dynamic control of protein secretion. The systems can be devised to retain or pass proteins of interest by using different molecular weight cutoff membranes for the filters and different configurations of cell growth and medium circulation. Figure 11.4 shows a hollow fiber setup.

——— SPECIAL SUBSTRATES FOR CELL CULTURE ———

GROWTH IN SEMISOLID MEDIA

Cells may be grown in semisolid media to prevent attachment. The ability for single cells to grow in semisolid medium is often used as an indicator of transformation from a normal phenotype to a tumorigenic phenotype. There are, however, some normal cells that

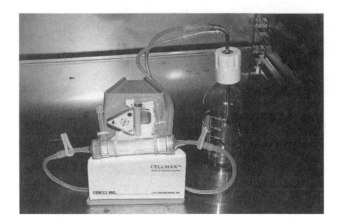

Figure 11.4. Hollow fiber systems for cell culture (Millipore setup).

can grow, especially if plated in small aggregates, in semisolid media. Here, the lack of attachment maintains a more rounded 3-dimensional shape *in vitro* that can help maintain differentiated cell function in some cell types (Halban *et al.,* 1987; Verhoeven *et al.,* 1986).

Materials

1. Difco Noble agar
2. 2× growth medium
3. Purified water
4. Water bath, set at 45°C
5. 15-ml conical tubes
6. 125-ml glass bottle with screw cap
7. 60-mm plastic petri dishes

Procedure

1. Add 1 g agar to 100 ml purified water in a glass bottle. Loosely cap. This is the 1% agar solution.
2. Add 0.5 g agar to 100 ml purified water in a glass bottle. Loosely cap. This is the 0.5% agar solution.
3. Autoclave for 30 min at 120°C.
4. Transfer the bottles to a 45°C water bath.
5. Prepare the 0.5% bottom agar layer by adding 2 ml of 1% agar to 2 ml 2× growth medium in a 15-ml conical tube.
6. Resuspend and add to a 60-mm plate, taking care not to introduce air bubbles that might create holes in the agar. Set aside and allow time to gel at room temperature (about 30 min).
7. Trypsinize, neutralize, and count cells to be plated. Prepare the cell suspension so that the seeding density/dish is in a 100-μl volume.
8. Prepare the 0.25% top agar layer by adding 1.5 ml of 1% agar to 1.5 ml 2× growth medium in a 15-ml conical tube.
9. Add 100 μl of the cell suspension to each 15-ml conical tube of cooled (40°C) agar.

10. Resuspend and add to each previously prepared dish.
11. Swirl the dish gently to allow even spreading.
12. Place in a 37°C, 5% CO_2 humidified incubator and leave for 7–10 days.

COLLAGEN GELS

Collagen gels also can be used to maintain cells in a more rounded, less spread configuration. While these are often more expensive and difficult to prepare than semisolid media, they do provide a more physiologically relevant substrate and therefore may improve cell function. The gel also may be detached from the surface of the culture dish, allowing a floating raft to form for the cells to grow on. In this configuration, the cells receive their nutrients from the medium on their basal surface and are close to the oxygen-rich atmosphere on the top. The raft can also contract and change configuration with the cells. Cells, such as normal mammary cells (Stampfer and Yaswen, 1994) and endothelial cells (Nishida *et al.,* 1993), can re-create complex three-dimensional structures similar to their cell–cell association *in vivo* in these cultures.

Materials

1. Rat tail collagen (30 mg, Boehringer-Mannheim)
2. 0.2 M acetic acid
3. Neutralizing buffers:
   A. 12 mg/ml $NaHCO_3$ in 0.1 N NaOH
   B. 1.3 M NaCl in 0.2 M Na_2HPO_4
4. 24-well, 12-well, or 6-well plates
5. 50-ml conical tubes
6. Trypsinized, neutralized cell suspensions

Procedure

1. Reconstitute collagen in 12 ml 0.2 N acetic acid. Store at 4°C.
2. On ice, make a solution mixing 800 μl reconstituted collagen, with 100 μl neutralizing buffer A and 100 μl neutralizing buffer B, for each 1 ml of collagen solution. Mix well.
3. Add 250 μl to each well of a 24-well plate, 500 μl to each well of a 12-well plate, or 1 ml to each well of a 6-well plate. Allow to gel in a 37°C incubator (this takes around 20–30 min).
4. When the gel has set, add at least 1×10^5 cells/ml to the as yet ungelled collagen solution on ice. Resuspend well and add an equal volume of this solution to the collagen in the well.
5. If it is neither desired nor necessary to incorporate the cells into a gel, the cells may be resuspended in $1\times$ growth medium and added on top of the set gel. In this case the liquid volumes should be 500 μl/well for 24-well plates, 1 ml/well for 12-well plates, and 2 ml/well for 6-well plates.

Air–Liquid Interface

Collagen rafts can be made by suspending the cells in collagen at the time of plating. Then, once set, carefully detach the gel with a small sterile spatula. Lay the gel on a stainless steel grid that has been bent along the edge to keep the gel off the surface of the dish.

Small 1-mm or 2-mm-thick "rings" of silicon tubing can be cut and placed under the grid as well. Stainless steel sieves (in graduated mesh sizes) work well, can be sterilized, and are thin enough to cut to size if necessary.

CELL–CELL INTERACTION

Cells communicate with each other both *in vivo* and *in vitro*. *In vivo* cell–cell communication can be through direct cell–cell contact, through secretion of soluble factors that traverse short distances to act on another nearby cell (paracrine and autocrine effects), or through long-distance signaling through the body's circulatory or lymphatic systems (endocrine signaling). Work in the last two decades has repeatedly emphasized the importance of paracrine and autocrine pathways in regulating cell survival and function (King *et al.,* 1996; Mather, 1984; Mather *et al.,* 1992). These short-range interactions are extremely difficult to study *in vivo* because it is difficult to ablate such pleiotropic factors (for example, IGF-I and TGF-β) or provide them to only a single cell type in a specific organ. *In vitro* systems have contributed significantly to the discovery of these "growth factors" and to our understanding of their roles *in vivo*.

One method of studying the interactions of two cells *in vitro* is to directly culture the two cells in the same dish. Such direct coculture can provide useful data on the interactions of two distinct cell types. Examples of direct cocultures are shown in Figs. 11.5, 9.1, and 1.3. In the case of the cultures shown in Fig. 9.1B, transwell culture was effective in supporting cell growth (Li *et al.,* 1996). The formation of the folliclelike structure shown in Fig. 1.3, however, required direct contact of the two cell types involved (Li *et al.,* 1995). In contrast, complex, three-dimensional forms can also arise in monocultures such as the cap-

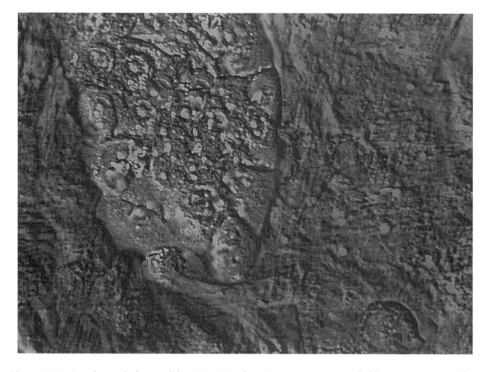

Figure 11.5. Coculture of a lung cell line (RL-65) colony (top center) surrounded by primary neonatal myocytes.

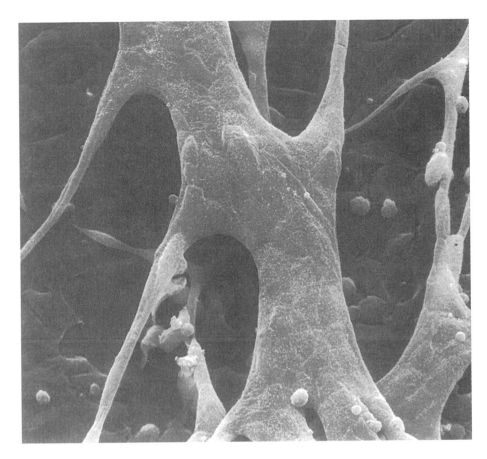

Figure 11.6. SEM of a culture of the TR-1 cell line. Cultures of cloned TR-1 capillary cell line form hollow branching tubelike structures that resemble capillaries *in vitro.*

illarylike tubular structures formed by the TR-1 capillary endothelial cell line (Fig. 11.6). The effect of the coculture of RL-65 lung cells with myocytes shown in Fig. 11.5 could be partially but not completely reproduced with conditioned medium from the RL-65 cells or purified hypertrophic factors (King *et al.,* 1996). We will discuss two methods that are refinements on the direct coculture of two cell types.

Feeder Layers

Cell feeder layers have been used for years to culture cells that can otherwise not be cultured alone. Feeder layers provide a special type of coculture environment in which a monolayer of one cell type (the feeder layer) is treated in such a manner that it can metabolize nutrients and produce macromolecules but is incapable of further cell division. Two such treatments are lethal irradiation (Puck and Marcus, 1995) and mitomycin C treatment. The second cell type is then cocultured on the feeder layer, which provides growth factors and/or an appropriate attachment surface. If a cell line is to be carried on feeder layers, it is important that the investigator have markers that can differentiate between the two cell types. Then, any cross-contamination, resulting from a few feeder cells escaping the treatment intact, can be detected. In many cases, the use of feeder layers can be avoided by switching to a richer nutrient mixture, providing attachment factors or extracellular matrix to the cells, supplementing serum-containing media with hormones or using a serum-free hormone-supplemented medium (see Chapter 8).

PROTOCOL TO CREATE FEEDER LAYERS USING MITOMYCIN C

Materials

1. Mitomycin C (MMC) solution: 5 mg/ml in PBS. Filter sterilize and store at 4°C in a light-tight container. Stable for about a week.
2. Confluent cell monolayer in desired plate size.

Procedure

1. To irreversibly inhibit proliferation, add 5 μg/ml MMC to the confluent dish of cells.
2. Incubate at 37°C for 2 hr.
3. Aspirate and wash plate with medium or PBS.
4. Trypsinize, neutralize, and split 1:3 or 1:4.
5. Plates may be stored at 37°C in normal growth medium for 24–48 hr before use.

TRANSWELL DISHES

Transwell dishes can be used when the investigator wishes to determine whether an observed effect of cell coculture is mediated via a soluble factor, to study directional secretion by cells, to study directional cell migration, and to observe the effect of exposing cells to different environments on their basal and luminal surface. Transwell inserts (Costar) come in 24-well and 6-well formats. The inserts are available with different membranes having different surface treatments and pore sizes. One can also apply the above-described collagen gels or other attachment factors to these inserts. Some of the membranes are transparent when wet, allowing observation of the cells through an inverted microscope; others are translucent and not good for direct observation of cells. Some membranes have pores large enough for cells to migrate through; others allow passage of macromolecules only. The membranes can be cut out of the transwells and the cells fixed and embedded and cut for electron microscopic observation (for example, see Fig. 3.7). Membranes can be cut and the cells used for biochemical analyses. The type of membrane selected will be determined by the experimental design. Table 3.1 lists different types of membranes that can be used to grow cells and their uses (see Chapter 3).

Materials

1. Costar transwell dishes, 6-well, with inserts
2. Cell suspension
3. Growth medium

Procedure

1. Prepare cells as described.
2. Carefully and aseptically remove the inserts, setting them down on the inside of the lid.
3. Add 1.5 ml growth medium to each well.

4. Replace the insert, making sure they are evenly sitting in the well, and add 1 ml of cells suspended (5×10^4/ml) in growth medium.
5. Adjust the insert so that the bottom is interfacing with the medium in the well. If necessary, add medium.
6. Incubate the cells for 24 hr, at 37°C.
7. Carefully aspirate the medium from each insert (if an air/medium interface is desired).
8. Monitor the plates daily and maintain the medium level in the well.

Transwell plate configurations are also useful for studying cells that secrete vectorially by comparing the levels of secreted protein in the well and the insert. These cultures have been used to study transport and electrically coupled cell monolayers. Inserts can also be used to determine whether direct cell–cell contact is involved in an observed phenomenon in coculture by setting up transwell cocultures with one cell type in the well and one in the insert, so that the two cell types are not in physical contact with each other. The wells with large pore sizes can be used to study factors that regulate cell migration through the pores from the upper chamber to the lower.

SUMMARY

The above methods are only a small sampling of the many complex culture systems that are used, from the highly miniaturized Terasaki and hanging drop cultures, which can be used to grow single cells, to the complex culture of several cell types used to mimic the various layers of the skin. Equipment to monitor cellular conductivity, pH, and calcium fluxes in living cultures have also been developed. It is clear that the possibilities for new techniques to grow cells will continue to increase with advances in technology and our understanding of the physiology and biochemistry of cell functioning *in vitro*. This will, in turn, continue to increase our understanding of the complex interrelationships between cells *in vivo*.

REFERENCES

Buttke, T. M., McCubrey, J. A., and Owen, T. C., 1993, Use of an aqueous soluble tetrazolium/formazan assay to measure viability and proliferation of lymphokine-dependent cell lines, *J. Immunol. Methods* **157**:233–240.

Connolly, D. T., Knight, M. B., Harakas, N. K., Wittwer, A. J., and Feder, J., 1986, Determination of the number of endothelial cells in culture using an acid phosphatase assay, *Anal. Biochem.* **152**:136–140.

Halban, P., Powers, S. L., George, K. L., and Bonner-Weir, S., 1987, Spontaneous reassociation of dispersed adult rat pancreatic islet cells into aggregates with three-dimensional architecture typical of native islets, *Diabetes* **36**:783–790.

King, K., Winer, J., and Mather, J., 1996, Endogenous cardiac vasoactive factors modulate endothelin production by cardiac fibroblasts in culture, *Endocrine* **5**:95–102.

Li, R., Phillips, D. M., and Mather, J. P., 1995, Activin promotes ovarian follicle development *in vitro*, *Endocrinology* **136**:849–856.

Li, R. H., Gao, W.-Q., and Mather, J. P., 1996, Multiple factors control the proliferation and differentiation of rat early embryonic (day 9) neuroepithelial cells, *Endocrine* **15**:205–217.

Mather, J. P., 1984, Intratesticular regulation: Evidence for autocrine and paracrine control of testicular function, in: *Mammalian Cell Culture: The Use of Serum-Free and Hormone Supplemented Media* (J. P. Mather, ed.), Plenum Press, New York, pp. 167–194.

Mather, J. P., Woodruff, T. K., and Krummen, L. A., 1992, Paracrine regulation of reproductive function by inhibin and activin, *Proc. Soc. Exp. Biol. Med.* **201**:1–15.

Nishida, M., Carley, W. W., Gerritsen, M. E., Ellingsen, O., Kelly, R. A., and Smith, T. W., 1993, Isolation and characterization of human and rat cardiac microvascular endothelial cells, *Am. J. Physiol.* **246:**H639–H652.

Puck, T., and Marcus, P., 1955, A rapid method for viable cell titration and clone production with HeLa cells in tissue culture: The use of x-irradiated cells to supply conditioning factors, *Proc. Natl. Acad. Sci. USA* **41:**432–437.

Roehm, N. W., Rodgers, G. H., Hatfield, S. M., and Glasebrook, A. L., 1991, An improved colorimetric assay for cell proliferation and viability utilizing the tetrazolium salt XTT, *J. Immunol. Methods* **142:**257–265.

Scudiero, D. A., Shoemaker, R. H., Paull, K. D., Monks, A., Tierney, S., Notziger, T. H., Currens, M. T., Seniff, D., and Boyd, M. K., 1988, Evaluation of a soluble tetrazoilium/formazan assay for cell growth and drug sensitivity in culture using human and other tumor cell lines, *Cancer Res.* **48:**4827–4833.

Stampfer, M. R., and Yaswen, P., 1994, Growth, differentiation, and transformation of human mammary epithelial cells in culture, *Cancer Treat. Res.* **71:**29–48.

Verhoeven, G., Cailleau, J., Van der Schueren, B., and Cassiman, J. J., 1986, The dynamics of steroid and adenosine 3′,5′-cyclic monophosphate output in perfused interstitial cell aggregates derived from prepubertal rat testis, *Endocrinology* **119:**1476–1488.

CHAPTER 12

Cell Culture for Commercial Settings

In the last decade, cell culture has come into its own as an area of considerable commercial importance in many biotechnology and pharmaceutical companies (Arathoon and Birch, 1986). The previous chapters have focused on techniques needed in the research laboratory, whether it is located in an academic or commercial setting. These techniques all apply to the cell culture technology that is part of the discovery of new products in biotechnology, whether it is expression cloning of a protein, purification of a desired activity using an *in vitro* bioassay, the production of a transgenic mouse using embryonic stem (ES) cells, or the optimization of new vectors for mammalian cell expression of recombinant proteins. This chapter will detail some special considerations that apply to cell culture as performed in commercial settings. As with previous chapters on specialized tissue culture techniques, we will not attempt to give complete details on techniques used in this setting, but will refer to published material describing techniques in each subdiscipline.

Generally speaking, the constraints on commercial tissue culture fall into three categories: (1) Speed is of essence; (2) processes must be amenable to scale up either to a rapid, automated, high-throughput format for bioassays or to large-volume production for recombinant proteins; and (3) the process and culture methods used must meet the standards of documentation, control, and reproducibility required for regulatory approval. The most efficient and successful development scientists keep these goals in mind from the beginning of the task of developing these processes and the cell lines used in them.

In this chapter we will attempt to introduce the student or scientist to a general description of how the aims of commercial cell culture can differ from common practice in a research or teaching laboratory setting. Because of commercially held trade secrets and the highly individualized nature of these processes, a complete mastery of commercial cell culture can only be gained through actual experience in a commercial setting. It should be emphasized, however, that the basic principles that control cell replication and function are the same at any scale and in any setting. The problems of adapting cell culture to commercial settings therefore is one of understanding the difference in goals (e.g., maximum titer rather than modeling of *in vivo* responses) and the different environments in which the cells are expected to function. The scientific principles the industrial cell biologists use to accomplish these goals are the same as those outlined throughout this book, although the equipment used may vary.

—— THE CELL AS INDUSTRIAL PROPERTY ——

The cells expressing a commercial recombinant protein product become a valuable asset. Newly developed or genetically modified cell lines, as well as newly developed media, may be patentable to protect the investment made in developing such tools. In addition, regulatory agencies approve the use of a specific cell line at specified passage numbers for production of recombinant or natural protein products. Thus, the description, characterization, frozen storage, and thaw of the production cell line is critical to the continued success of the product from initial development through the life of the marketed product.

ENGINEERING CELLS FOR SPECIFIC PROPERTIES

One great advantage of cells as living factories is the ability to select or engineer cells to express specific characteristics. This is possible even without a complete understanding of the biological pathways involved in the alterations. One familiar example is adapting cells to suspension growth. We still do not understand completely the role of substrate in regulating cell function, although there is a growing awareness of its importance and the complexity of the processes involved. Nevertheless, it is possible to select cells that will lose their requirement for attachment and grow in suspension, using procedures such as those outlined in Chapter 11. Similarly, one can adapt cells to grow in lower serum concentrations, to higher densities, and so forth. Thus, the cell may be selected to have the characteristics most desirable for use in a commercial setting. The degree to which this alters the cell's properties from those of its *in vivo* counterpart are important only to the extent, if any, that these alterations affect the final product to be produced. Thus, one can use the adaptability of living cells to their environment to save much time and effort in optimizing a cell culture process.

Additionally, molecular biology techniques can be used to engineer cells that have the properties desirable in a production cell. One example is the transfection of cells to allow the production of required growth factors, such as insulin, which then may act in an autocrine fashion during production runs. This leads to a less expensive and more robust culture system, since the required protein is being produced continuously in the culture. An example of this approach is shown in Fig. 12.1. In this instance, insulin was known to be required for optimal growth and secretion of protein by CHO cells. Insulin is normally made and processed from proinsulin by pancreatic beta cells. Very few cells, including CHO, can correctly process proinsulin. Therefore, in order to provide an effective autocrine growth environment, the proinsulin either must be processed or must be active without processing. The latter is the case for CHO. Therefore, cells transfected with and selected for stable expression of the proinsulin gene were able to grow as well in the absence of exogenous insulin as the parent line grew with the addition of optimal insulin (Fig. 12.1) (Mather and Ullrich, 1988). Thus, cells can be intentionally genetically altered to improve their characteristics as production cell lines. However, when using this approach, as with selection methods, care must be taken not to sacrifice optimal growth characteristics or the quality of the product when optimizing the cells for expression of a specific gene.

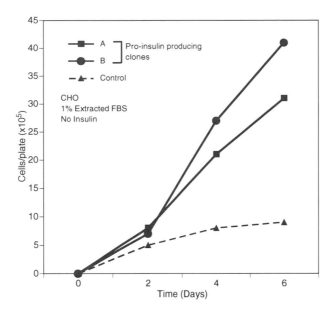

Figure 12.1. Growth of two CHO clones transfected with the proinsulin gene and selected for insulin independent growth and the CHO parent cells in the absence of externally added insulin. The proinsulin-secreting clones grow well in the absence of any added insulin.

PREPARING, CHARACTERIZING, AND STORING CELL BANKS

The cell bank of the production cell line for a recombinant commercial product is an invaluable resource. The ability to continue making the identical product over several decades depends on its integrity. A good cell bank, proven to be free of adventitious agents, with high viability from thaw and good growth characteristics, is crucially important to the ultimate regulatory approval and commercial success of a recombinant protein product. Again, the considerations for successful cell banking are essentially the same whether it is done on the laboratory scale, as described in Chapter 5, or on an industrial scale, but the equipment used and degree of testing performed on the banked cells will vary.

Generally speaking, two types of cell banks may be used to preserve a production cell line: a master cell bank and a working cell bank. The master bank is the completely characterized bank from which all cells used in commercial production must derive. This bank consists of from 100 to 500 frozen ampoules derived from the same culture and passage. The bank is put down in permanently labeled glass ampoules, which are then leak tested by immersion in a liquid dye. The vials are frozen using an automated freezing machine that lowers the temperature at the required rate for the entire set of vials at once. The master cell bank must be fully characterized as to identity and freedom from adventitious agents, including viruses. Such a complete characterization may take many weeks and cost several hundred thousand dollars. Vials from the master cell bank are generally stored at more than one site.

The working banks are laid down periodically by thawing a vial of the master bank, expanding the cells over a designated number of passages, and refreezing the cells. The working bank characterization need not be as extensive as the master bank characterization, although sterility and identity must be verified. The cells are then thawed and expanded for the actual production cultures. This allows more flexibility in storage and transport of the bank since all banked cells do not have to be laid down at once. Cells are used for production only over a specified number of passages. The passage number should be long enough to allow the cells to be expanded from the frozen vial and used for several production runs before they must be discarded (Lubiniecki, 1990). The cells and product must not vary in their characteristics during this period. It is obvious that there is a strong advantage to having stable, clonal cell cultures as production lines. The periodic replacement of the cells with new cultures thawed from the banks minimizes the chance of undetected changes occurring in the cultures.

VERY SMALL SCALE

Bioassays are becoming increasingly important both as a research tool and as a required quality control tool for commercial production of recombinant proteins. The raw materials used in large-scale fermentation processes must be lot tested as they are purchased. Since the purpose of these media and media additives is to support cell growth, growth should be used as a quality control test. The bioactivity of the end product can also be tested *in vitro* in many instances. While some products must still be tested *in vivo,* the desire to reduce animal usage and increasing data on the reliability of *in vitro* testing will undoubtedly lead to an increased use of *in vitro* tests for bioactivity for product release.

Product discovery can also benefit greatly from high-throughput assay screens. A given biological activity may be defined (e.g., growth stimulation of endothelial cells). This activity may then be followed through a conventional biochemical purification scheme (Ferrara *et al.,* 1991), or during an attempt to expression clone the gene responsible for the activity (Pennica *et al.,* 1996). In both instances, high-throughput assays, such as those described in Chapter 11, will reduce the cost and labor required.

The rise of genomics and computerized databases and the availability of cloned sequences of many proteins as well as combinatorial libraries of small molecules increase the importance of the use of high-throughput *in vitro* bioassays for rapid screening of these proteins to first identify biologically active compounds and then to further define their biological functions.

VERY LARGE SCALE

Another growing application of cell culture in an industrial setting is the production of large quantities of recombinant proteins for pharmaceutical or other industrial uses. While the biotechnical industry was initially based on the use of bacterial production systems, it was recognized some years ago that many useful proteins could not be produced in their active form in bacteria, or that the yields from protein purification and refolding, often a few percent of the total, would be so low as to make a mammalian cell culture process more economical. These include proteins that require appropriate glycosylation for activity and those that are very difficult to refold such as proteins containing multiple disulfide bonds. In addition, the new technologies to "humanize" mouse monoclonal antibodies (Werther *et al.,* 1996) or clone human antibodies directly from phage antibody libraries (Portalano *et al.,* 1993) using molecular biology techniques make it, in some instances, easier to produce the desired antibodies through recombinant expression techniques than in hy-

Figure 12.2. Very large-scale culture: Filling a 12,500-liter cell culture fermenter with medium.

bridomas. Alternatively, large-scale production of antibodies from hybridomas for medical or industrial use (e.g., in protein purification) requires cell culture expertise as well.

In this book we have previously dealt with only methods relating to procedures that can be accomplished in an ordinary laboratory. In this type of laboratory setting, if larger-scale cell culture is desired, the use of roller bottles or spinners (see Chapter 11) will generally be sufficient to meet the needs. These can be scaled to tens of liters of culture medium with a minimum investment in equipment and personnel training. However, in the industrial production of proteins for pharmaceutical or other markets, this scale is frequently not adequate to meet the demands of supply and cost. These industrial processes may need to produce kilogram quantities of highly purified protein per year in order to meet market needs. Industrial-scale fermenters can be as large as 12,000 liter volumes. Figure 12.2 shows the viewport of a 12,000-liter mammalian cell culture fermenter being filled with sterile culture medium. Both stirred-tank and air-lift fermenters have been used for large-scale culture (Arathoon and Birch, 1986). Designing the equipment and the processes for these very large-scale plants and operating the plants require the close collaboration of cell biologists and engineers.

One method of scaling-up a process is to take a laboratory-scale process, such as production in roller bottles, and just do more of it. Robotics can be used to facilitate repetitive tasks. However, when very large quantities of protein are required, this method is generally not cost-effective and has a limited potential for cost savings. An alternative, outlined in Fig. 12.3, is to adapt laboratory processes, such as growth in spinners, that can be run in equipment especially designed to be scalable at very large scales, such as fermenters. The initial optimization can be performed in small-scale plate culture, but cells will sometimes perform differently in fermenters. Therefore, data obtained in plates should be checked first in small minifermenters, such as that shown in Fig. 12.4. The process can then be scaled up to the desired level using larger fermenters such as those shown in Fig. 12.5. If care is taken

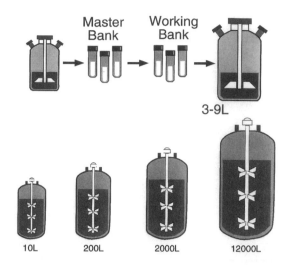

Figure 12.3. Schematic of the process of scaling up a cell culture process.

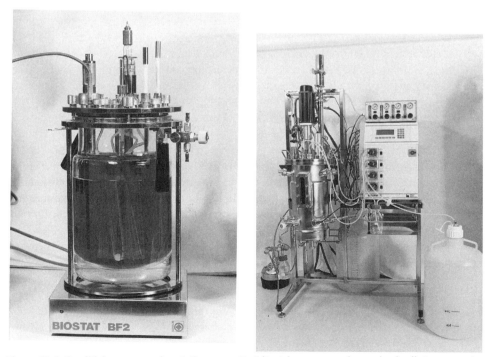

Figure 12.4. Small laboratory-scale minifermenter (1–2 liter) for automated growth of cells in suspension. The larger fermenter on the right might be used in a pilot scale facility. (Courtesy of B. Braun Biotech Inc.)

Figure 12.5. Large production fermenters (1000 liter) such as might be used in commercial production of a recombinant protein or antibody from mammalian cell cultures. (Courtesy of Pharmacia and Upjohn. Cell culture plant of Pharmacia and Upjohn; from B. Braun Biotech Inc.)

to design the smaller-scale cultures to closely follow the large-scale conditions, recombinant protein yields are similar from cells cultured over a wide range of equipment (Fig. 12.6). Here, the expense of the equipment, which may not be justified for laboratory-scale experiments, is offset by the considerable cost savings realized when the process is scaled up.

Much of the cost of producing pharmaceutical-grade proteins lies in the extensive testing that must be carried out to prove that each lot of protein is pure, free of contaminant proteins or other agents, and active. This testing requires the same amount of time and material whether the lot is 5 g of material or 5 kg of material. Obviously, the bigger the lot, the better. Figure 12.7 shows a large-scale column used for industrial-scale protein purification. Additionally, large-scale facilities must be staffed 24 hr a day to monitor cell culture runs and operate equipment. Obviously the labor required to produce a given amount of protein decreases with increasing scale.

While the basic principles of cell culture remain the same at all scales, there are special considerations that arise when doing large-scale culture (Mather and Moore, 1997; Mather, 1990). The large scale, of necessity, raises engineering problems that require the addition of engineers to the process development team. The entire process should be seen as a whole, and all the groups involved should collaborate from the beginning. Figure 12.8 diagrams some of the interactions necessary in designing a process. If the molecular biologist who designs the vectors, the cell biologists who design the medium, the engineers who design the fermentation equipment, and the protein chemists who purify the product work independently, the resulting process is likely to be awkward and pieced together, with a

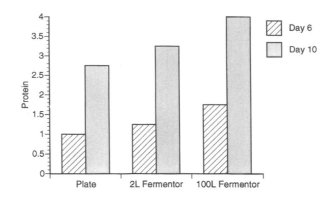

Figure 12.6. Comparison of protein yields of cells grown at different scales. While there is some difference in the yield at different scales, the increase with time is proportional at all scales.

Figure 12.7. A protein purification column for large-scale purification.

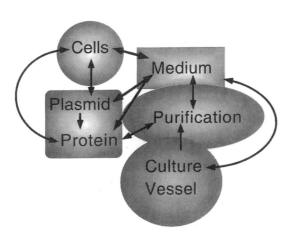

Figure 12.8. Interacting variables contributing to the creation of a large-scale industrial cell culture process. It is obvious that most of the "compartments" interrelate with other aspects of the process. It should therefore be designed as a whole for optimal performance.

higher failure rate and higher cost compared to a process designed with input from all contributors and an understanding of how the variables of the cell culture and purification processes interrelate to determine yield, reliability, cost, and protein quality.

In industrial-scale culture, considerations relating to the reproducibility, cost, and safety of the process are of primary concern. A major goal is to achieve as reproducible a process as possible. If one run shows too much unexplained variance from the norm, the product must be discarded. This reproducibility, which is at a level far beyond what is usually achieved in the ordinary research laboratory setting, can only be obtained with a well-understood cell culture system, well-trained personnel, and a rigorously controlled process. Clearly, if the goal is to run exactly the same culture conditions for hundreds or thousands of repeats, time spent optimizing and simplifying the process to minimize cost and complexity will pay off. The cells selected should be as robust as possible, with minimal requirements for complex growth conditions or expensive medium additives. The cells used in large-scale industrial culture are used for only a strictly controlled number of passages after thaw from a well-characterized cell bank. Therefore, they should have a high viability and optimal growth rate immediately on thawing. This can be achieved by optimizing the freezing and thawing conditions and freezing the cells, in relatively large aliquots, in the same medium they will be expected to grow in on thaw (with the addition of the freezing agent). See Chapter 5 for details on freezing and thawing cells.

Medium will generally be prepared, sterilized, and used immediately, since space for storage would be prohibitive. The medium used to grow the cells must be as inexpensive as possible. Some components of the medium may be substituted for by others that are less expensive or previously approved for use in human pharmaceuticals. Thus, spending the time to optimize the nutrient mixture and other components of the medium will pay off over the lifetime of the process. Just the physical constraints of very large-volume cell culture can be an issue. Adding 10 μl of an insulin solution to a culture dish containing 10 ml of medium is physically undemanding. However, adding the equivalent amount of insulin to a 12,000-liter tank requires mixing, sterilizing, and adding 12 liters of solution.

Since the cells in fermenters cannot be easily observed visually, there is a complex network of equipment for monitoring and regulating the temperature, pH, PO$_2$, mixing, and so forth. This allows a higher degree of control and reproducibility than can be obtained easily using an incubator and tissue culture plates. However, the equipment itself can introduce problems, such as shear or bubbles from aeration, which can effect cell viability and must be dealt with. The equipment used is complex and must be monitored 24 hr a day to make sure everything is running smoothly. The process designed also should require a minimum of handling and additions so as to minimize the chance of introducing a contaminant into the cultures. Even disposal of the cells after harvest can present a problem when there are 10^{10} cells to dispose of. Since mammalian cell culture at this scale has been performed for little more than a decade, there is still plenty of opportunity for continuing innovation in this area of culture.

REFERENCES

Arathoon, W. R., and Birch, J. R., 1986, Large-scale cell culture in biotechnology, *Science* **232:**1390–1395.

Ferrara, N., Leung, D., Cachianes, G., Winer, J., and Henzel, W., 1991, Purification and cloning of vascular endothelial growth factor secreted by pituitary folliculostellate cells, *Methods Enzymol.* **198:**391–405.

Goeddel, D. V. (ed.), 1991, *Methods in Enzymology,* Vol. 185, Academic Press, San Diego.

Lubiniecki, A., 1990, Continuous cell substrate considerations, in: *Large-Scale Mammalian Cell Culture Technology* (A. Lubiniecki, ed.), Marcel Dekker, New York, pp. 495–513.

Mather, J. P., 1990, Optimizing the cell and culture environment for the production of recombinant proteins, *Methods Enzymol.* 185:157–167.

Mather, J., and Moore, A., 1998. Culture media: Large scale production of proteins in animal cells, in: *The Encyclopedia of Bioprocess Technology: Fermentation, Biocatalysis, & Bioseparation* (M. Flickinger and S. Drew, eds.), John Wiley & Sons, New York, in press.

Mather, J., and Ullrich, A., 1988, Culturing recombinant host cells by first transforming host cell with nucleic acid encoding polypeptide factor necessary for growth, European Patent # EP 307247.

Pennica, D., Wood, W., and Chien, K., 1996, Cardiotropin-1: A multifunctional cytokine that signals via LIF receptor-gp 130 dependent pathways, *Cytokine Growth Factor Rev.* **7:**81–91.

Portalano, S., McLachlan, S., and Rapoport, B., 1993, High affinity, thyroid specific human autoantibodies displayed on the surface of filamentous phage use V genes similar to other autoantibodies, *J. Immunol.* **151:**2839–2851.

Werther, W., Gonzalez, T., O'Connor, S., McCabe, S., Chan, B., 1996, Humanization of an anti-lymphocyte function-associated antigen (LFA)-1 monoclonal antibody and reengineering of the humanized antibody for binding to rhesus LFA-1, *J. Immunol.* **157:**4986–4995.

Glossary

Adventitious   Developing from unusual points of origin, such as embryos from sources other than zygotes. This term can also be used to describe agents that contaminate cell cultures.

Anchorage-dependent cells or cultures   Cells or cultures derived from them that will grow, survive, or maintain function only when attached to a surface such as glass or plastic. The use of this term does not imply that the cells are normal or that they are or are not neoplastically transformed.

Aneuploid   The situation that exists when the nucleus of a cell does not contain an exact multiple of the haploid number of chromosomes, one or more chromosomes being present in a greater or a lesser number than the rest. The chromosomes may or may not show rearrangements.

Asepsis   Without infection or contaminating microorganisms.

Aseptic technique   Procedures used to prevent the introduction of fungi, bacteria, viruses, mycoplasma, or other microorganisms into cell, tissue, and organ culture. Although these procedures are used to prevent microbial contamination of cultures, they also prevent cross-contamination of cell cultures as well. These procedures may or may not exclude the introduction of infectious molecules.

Attachment efficiency   The percentage of cells plated (seeded, inoculated) that attach to the surface of the culture vessel within a specified period of time. The conditions under which such a determination is made should always be stated.

Autocrine cell   In animals, a cell that produces hormones, growth factors, or other signaling substances for which it also expresses the corresponding receptors. See also *endocrine* and *paracrine*.

Axenic culture   A culture without foreign or undesired life forms. An axenic culture may include the purposeful cocultivation of different types of cells, tissues, or organisms.

Cell bank*   A number of vials of frozen cells that have been derived from a single, well-characterized cell culture and maintained at liquid nitrogen temperatures. Over a pe-

The definitions have been excerpted from "Terminology Associated with Cell, Tissue and Organ Culture, Molecular Biology and Molecular Genetics," *In Vitro Cell. Dev. Biol.* **26**:97–101, 1990, with permission of the Tissue Culture Association. Definitions marked with an asterisk (*) are those of the authors.

riod of years, cells can be thawed and grown from the bank that should have the same characteristics as the original culture from which the bank was prepared.

Cell culture   Term used to denote the maintenance or cultivation of cells *in vitro,* including the culture of single cells. In cell cultures, the cells are no longer organized into tissues.

Cell generation time   The interval between consecutive divisions of a cell. This interval can best be determined at present with the aid of cinephotomicrography. *This term is not synonymous with population doubling time.*

Cell hybridization   The fusion of two or more dissimilar cells leading to the formation of a synkaryon.

Cell line   A cell line arises from a primary culture at the time of the first successful subculture. The term *cell line* implies that cultures from it consist of lineages of cells originally present in the primary culture. The terms *finite* or *continuous* are used as prefixes if the status of the culture is known. If not, the term *line* will suffice. The term *continuous line* replaces the term *established line.* In any published description of a culture, one must make every attempt to publish the characterization or history of the culture. If such has already been published, a reference to the original publication must be made. In obtaining a culture from another laboratory, the proper designation of the culture, as originally named and described, must be maintained and any deviations in cultivation from the original must be reported in any publication.

Cell strain   A cell strain is derived either from a primary culture or a cell line by the selection or cloning of cells having specific properties or markers. In describing a cell strain, its specific features must be defined. The terms *finite* or *continuous* are to be used as prefixes if the status of the culture is known. If not, the term *strain* will suffice. In any published description of a cell strain, one must make every attempt to publish the characterization or history of the strain. If such has already been published, a reference to the original publication must be made. In obtaining a culture from another laboratory, the proper designation of the culture, as originally named and described, must be maintained and any deviations in cultivation from the original must be reported in any publication.

Chemically defined medium*   A nutritive solution for culturing cells in which each component is specifiable and ideally is of known chemical structure. Growth factors may be included in defined media if they are of known composition and purity (preferably synthetic or recombinant). Serum-free medium containing growth-promoting substances of unknown composition such as bovine pituitary extract or embryo extract should not be characterized as chemically defined.

Clone   In animal cell culture terminology, a population of cells derived from a single cell by mitoses. A clone is not necessarily homogeneous; therefore, the terms *clone* and *cloned* do not indicate homogeneity in a cell population, genetic or otherwise. In plant culture terminology, the term may refer to a culture derived as above or it may refer to a group of plants propagated only by vegetative and asexual means, all members of which have been derived by repeated propagation from a single individual.

Cloning efficiency   The percentage of cells plated (seeded, inoculated) that form a clone. One must be certain that the colonies formed arose from single cells in order to properly use this term. See also *colony-forming efficiency.*

Colony-forming efficiency   The percentage of cells plated (seeded, inoculated) that form a colony.

Complementation   The ability of two different genetic defects to compensate for one another.

Contact inhibition of locomotion   A phenomenon characterizing certain cells in which two cells meet, locomotory activity diminishes, and the forward motion of one cell over the surface of the other is stopped.

Continuous cell culture   A culture that is apparently capable of an unlimited number of population doublings; often referred to as an *immortal cell culture*. Such cells may or may not express the characteristics of *in vitro* neoplastic or malignant transformation. See also *immortalization*.

Crisis   A stage of the *in vitro* transformation of cells. It is characterized by reduced proliferation of the culture, abnormal mitotic figures, detachment of cells from the culture substrate, and the formation of multinucleated or giant cells. During this massive cultural degeneration, a small number of colonies usually, but not always, survive and give rise to a culture with an apparent unlimited *in vitro* life span. This process was first described in human cells following infection with an oncogenic virus (SV40). See also *cell line, in vitro transformation,* and *in vitro senescence.*

Cryopreservation   Ultralow temperature storage of cells, tissues, embryos or seeds. This storage is usually carried out using temperatures below $-100°C$.

Cumulative population doublings   See *population doubling level.*

Cybrid   The viable cell resulting from the fusion of a cytoplast with a whole cell, thus creating a cytoplasmic hybrid.

Cytoplast   The intact cytoplasm remaining following the enucleation of a cell.

Cytoplasmic hybrid   Synonymous with *cybrid.*

Cytoplasmic inheritance   Inheritance attributable to extranuclear genes, for example, genes in cytoplasmic organelles, such as mitochondria or chloroplasts, or in plasmids, and so forth.

Density-dependent inhibition of growth   Mitotic inhibition correlated with increased cell density.

Differentiated   Cells that maintain, in culture, all or much of the specialized structure and function typical of the cell type *in vivo.*

Diploid   The state of the cell in which all chromosomes, except sex chromosomes, are two in number and are structurally identical with those of the species from which the culture was derived.

Electroporation   Creation, by means of an electrical current, of transient pores in the plasmalemma, usually for the purpose of introducing exogenous material, especially DNA, from the medium.

Embryo culture   *In vitro* development or maintenance of isolated mature or immature embryos.

Embryogenesis   The process of embryo initiation and development.

Endocrine cell   In animals, a cell that produces hormones, growth factors, or other signaling substances for which target cells, expressing the corresponding receptors, are located at a distance. See also *autocrine* and *paracrine.*

Epigenetic event   Any change in a phenotype that does not result from an alteration in DNA sequence. This change may be stable and heritable and includes alteration in DNA methylation, transcriptional activation, translational control, and posttranslational modifications.

Epigenetic variation   Phenotypic variability that has a nongenetic basis.

Epithelial-like   Resembling or characteristic of, or having the form or appearance of epithelial cells. In order to define a cell as an epithelial cell, it must possess characteristics typical of epithelial cells. Often one can be certain of the histological origin and/or function of the cells placed into culture and, under these conditions, one can be reasonably confident in designating the cells as epithelial. It is incumbent upon the individual reporting on such cells to use as many parameters as possible in assigning this term to a culture. Until such time as a rigorous definition is possible, it would be most correct to use the term *epithelial-like*.

Euploid   The situation that exists when the nucleus of a cell contains exact multiples of the haploid number of chromosomes.

Explant   Tissue taken from its original site and transferred to an artificial medium for growth or maintenance.

Explant culture   The maintenance or growth of an explant in culture.

Feeder layer   A layer of cells (usually lethally irradiated for animal cell culture) on which are cultured a fastidious cell type. See also *nurse culture.*

Fermenter*   An apparatus for carrying out and regulating the growth of cells in suspension. Fermenters outfitted for use with mammalian cells will have automatic measurement and control of PO_2, pH, and temperature. The configuration of the tanks, propellers, and so forth, used for mammalian cells will differ from those used for yeast or bacterial fermentations.

Fibroblastlike   Resembling or characteristic of having the form or appearance of fibroblast cells. In order to define a cell as a fibroblast cell, it must possess characteristics typical of fibroblast cells. Often, one can be certain of the histological origin and/or function of the cells placed into culture and, under these conditions, one can be reasonably confident in designating the cells as fibroblast. It is incumbent upon the individual reporting on such cells to use as many parameters as possible in assigning this term to a culture. Until such time as a rigorous definition is possible, it would be most correct to use the term *fibroblastlike.*

Finite cell culture   A culture that is capable of only a limited number of population doublings after which the culture ceases proliferation. See *in vitro senescence.*

Heterokaryon   A cell possessing two or more genetically different nuclei in a common cytoplasm, usually derived as a result of cell-to-cell fusion.

Heteroploid   The term given to a cell culture when the cells comprising the culture possess nuclei containing chromosome numbers other than the diploid number. This is a term used only to describe a culture and is not used to describe individual cells. Thus, a heteroploid culture would be one that contains aneuploid cells.

Histiotypic   The *in vitro* resemblance of cells in culture to a tissue in form or function or both. For example, a suspension of fibroblastlike cells may secrete a glycosaminoglycan–collagen matrix and the result is a structure resembling fibrous connective tissue, which is therefore histiotypic. This term is not meant to be used along with the word *culture*. Thus, a tissue culture system demonstrating form and function typical of cells *in vivo* would be said to be histiotypic.

Homokaryon   A cell possessing two or more genetically identical nuclei in a common cytoplasm, derived as a result of cell-to-cell fusion.

Hybrid cell   The term used to describe the mononucleate cell that results from the fusion of two different cells, leading to a formation of a synkaryon.

Hybridoma   The cell that results from the fusion of an antibody-producing tumor cell (myeloma) and an antigenically stimulated normal plasma cell. Such cells are constructed because they produce a single antibody directed against the antigen epitope that stimulated the plasma cell. This antibody is referred to as a monoclonal antibody.

Hyperplasia*   An increase in cell number in a tissue or culture.

Hypertrophy*   An increase in cell size (volume). Hypertrophy results in an increase in protein–DNA ratio in the cells and can result in an increase in the surface area of cells in culture. A change in cell surface due to spreading without an increase in protein–cell is not hypertrophy.

Immortalization   The attainment of a finite cell culture, whether by perturbation or intrinsically, of the attributes of a continuous cell line. An immortalized cell is not necessarily one that is neoplastically or malignantly transformed.

Immortal cell culture   See *continuous cell culture.*

Induction   Initiation of a structure, organ, or process *in vitro.*

***In vitro* neoplastic transformation**   The acquisition by cultured cells of the property to form neoplasms, benign or malignant, when inoculated into animals. Many transformed cell populations that arise *in vitro* intrinsically, or through deliberate manipulation by the investigator, produce only benign tumors, which show no local invasion or metastasis following animal inoculation. If there is supporting evidence, the term *in vitro malignant neoplastic transformation* or *in vitro malignant transformation* can be used to indicate that an injected cell line does indeed invade or metastasize.

In vitro* senescence   In vertebrate cell cultures, especially human-derived cultures, the property attributable to finite cell cultures, namely, their inability to grow beyond a finite number of population doublings. Neither invertebrate nor plant cell cultures exhibit this property. Rodent cell lines established in serum-free conditions also frequently do not exhibit senescence.

***In vitro* transformation**   A heritable change, occurring in cells in culture, either intrinsically or from treatment with chemical carcinogens, oncogenic viruses, irradiation, transfection with oncogenes, and so forth, and leading to the acquisition of altered morphological, antigenic, neoplastic, proliferative, or other properties. This expression is distinguished from *in vitro neoplastic transformations* in that the alterations occurring in the cell population may not always include the ability of the cells to produce tumors in appropriate hosts. The type of transformation should always be specified in any description.

Karyoplast   A cell nucleus, obtained from the cell by enucleation, surrounded by a narrow rim of cytoplasm and a plasma membrane.

Liposome   A closed lipid vesicle surrounding an aqueous interior; may be used to encapsulate exogenous materials for ultimate delivery of these into cells by fusion with the cell.

Microcarriers*   Small, usually spherical, beads created to allow for mammalian cell attachment. These beads can be made of collagen, polystyrene, sepharose, glass, or other materials. Cells can grow on the exterior or interior surface of microcarriers. Microcarriers are generally used to allow the culture of attachment-dependent cells suspended in a liquid medium.

Microcarrier culture*   A culture in which cells can be grown in suspension by allowing the cells to attach to microcarriers, which are then suspended in medium and grown in spinners or fermenters.

Microcell   A cell fragment, containing one to a few chromosomes, which is formed by the enucleation or disruption of a micronucleated cell.

Micronucleated cell   A cell that has been mitotically arrested and in which small groups of chromosomes function as foci for the reassembly of the nuclear membrane, thus forming micronuclei, the maximum of which could be equal to the total number of chromosomes.

Mitogen*   A substance, hormone, or growth factor that induces cells to divide, leading to an increased cell number.

Morphogenesis   (1) The evolution of a structure from an undifferentiated to a differentiated state. (2) The process of growth and development of differentiated structures.

Morphogen*   A substance, hormone, or growth factor that induces morphogenesis.

Motogen* A substance, hormone, or growth factor that induces cell movement.

Mutant A phenotypic variant resulting from a changed or new gene.

Organ culture The maintenance or growth of organ primordia of the whole or parts of an organ *in vitro* in a way that may allow differentiation and preservation of the architecture and/or function.

Organized Arranged into definite structures.

Organogenesis The evolution, from dissociated cells, of a structure that shows natural organ form or function or both.

Organotypic Resembling an organ *in vivo* in three-dimensional form or function or both. For example, a rudimentary organ in culture may differentiate in an *organotypic* manner, or a population of dispersed cells may become rearranged into an *organotypic* structure and may also function in an organotypic manner. This term is not meant to be used along with the word *culture,* but is meant to be used as a descriptive term.

Paracrine In animals, a cell that produces hormones, growth factors, or other signaling substances for which the target cells, expressing the corresponding receptors, are located in its vicinity, or in a group adjacent to it. See also *autocrine* and *endocrine.*

Passage The transfer or transplantation of cell, with or without dilution, from one culture to another. It is understood that any time cells are transferred from one vessel to another, a certain portion of the cells may be lost, and therefore dilution of cells, whether deliberate or not, may occur. This term is synonymous with the term *subculture.*

Passage number The number of times the cells in the culture have been subcultured or passaged. In descriptions of this process, the ratio or dilution of the cells should be stated so that the relative cultural age can be ascertained.

Pathogen free Free from specific organisms based on scientific tests for the designated organisms.

Plant tissue culture The growth or maintenance of plant cells, tissues, organs, or whole plants *in vitro.*

Plating efficiency A term that originally encompassed the terms *attachment (seeding) efficiency, cloning efficiency,* and *colony-forming efficiency* and is now better described by using one or more of them in its place, as the term *plating* is not sufficiently descriptive of what is taking place. See *attachment, seeding, cloning,* and *colony-forming efficiency.*

Population density The number of cells per unit area or volume of a culture vessel. Also, the number of cells per unit volume of medium in a suspension culture.

Population doubling level The total number of population doublings of a cell line or strain since its initiation *in vitro.* A formula to use for the calculation of population doublings in a single passage is: Number of population doublings = $\text{Log}^{10}(N/No) \times 3.33$, where: N = number of cells in the growth vessel at the end of a period or growth; No = number of cells plated in the growth vessel. It is best to use the number of viable cells or number of attached cells for this determination. Population doubling level is synonymous with *cumulative* population doublings.

Population doubling time The interval, calculated during the logarithmic phase of growth, to which, for example, 1.0×10^6 cells increase to 2.0×10^6 cells. This term is not synonymous with *cell generation time.*

Primary culture A culture started from cells, tissues, or organs taken directly from organisms. A primary culture may be regarded as such until it is successfully subcultured for the first time. It then becomes a *cell line.*

Pseudodiploid This describes the condition where the number of chromosomes in a cell is diploid, but as a result of chromosomal rearrangements, the karyotype is abnormal and linkage relationships may be disrupted.

QS*   The process of adding sufficient liquid to a preexisting solution to bring the total to a stated volume, for example, "QS to 100 ml with water."

Recon or reconstituted cell   The viable cell reconstructed by the fusion of a karyoplast with a cytoplast.

Reculture   The process by which a cell monolayer or a plant explant is transferred, without subdivision, into fresh medium. See also *passage.*

Saturation density   The maximum cell number attainable, under specified culture conditions, in a culture vessel. The term is usually expressed as the number of cells per square centimeter in a monolayer culture or the number of cells per cubic centimeter in a suspension culture.

Seeding efficiency   See *attachment efficiency.*

Senescence   See *in vitro senescence.*

Somatic cell hybrid   The cell or plant resulting from the cell fusion of animal cells or plant protoplasts, respectively, derived from somatic cells that differ genetically.

Somatic cell genetics   The study of genetic phenomena of somatic cells. The cells under study are most often cells grown in culture.

Somatic cell hybridization   The *in vitro* fusion of animal cells or plant protoplasts derived from somatic cells that differ genetically.

Solara*   In large-scale culture, the process of moving cells from one vessel to another with a significant carryover of conditioned medium and a minimal addition of fresh medium.

Spinner*   A culture vessel for the growth of cells in suspension and maintenance of the cells in suspension. The vessel contains a suspended paddle that spins when in contact with a motorized base, thus mixing the medium and maintaining the cells in suspension. Spinners do not have instrumental control of temperature, pH, or PO_2 (see *fermenter*).

Sterile   (1) Without life. (2) Inability of an organism to produce functional gametes.

Subculture   See *passage.*

Substrain   A substrain can be derived from a strain by isolating a single cell or groups of cells having properties or markers not shared by all cells of the parent strain.

Surface- or substrate-dependent cells or cultures   See *anchorage-dependent cells.*

Suspension culture   A type of culture in which cells, or aggregates of cells, multiple while suspended in liquid medium.

Synkaryon   A hybrid cell that results from the fusion of the nuclei it carries.

Tissue culture   The maintenance or growth of tissues *in vitro* in a way that may allow differentiation and preservation of their architecture and/or function.

Totipotency   A cell characteristic in which the potential for forming all the cell types in the adult organism is retained.

Transfection   The transfer, for the purposes of genomic integration, of naked, foreign DNA into cells in culture. The traditional *microbiological* usage of this term implies that the DNA being transferred is derived from a virus. The definition as stated here is that which is in use to describe the general transfer of DNA irrespective of its source. See also *transformation.*

Transformation   See *in vitro transformation, in vitro neoplastic transformation,* and *transfection.*

Undifferentiated   With plant cells, a state of cell development characterized by isodiametric cell shape, with very little or no vacuole and a large nucleus, and exemplified by cells comprising an apical meristem or embryo. With animal cells, this is the state wherein the cell in culture lacks the specialized structure and/or function of the cell type *in vivo*.

Variant   A culture exhibiting a stable phenotypic change whether genetic or epigenetic in origin.

Virus-free   Free from specified viruses based on tests designed to detect the presence of the organisms in question.

Formula for Calculating Osmolarity

If one does not have an osmometer, the osmolarity of medium can be increased by a calculated amount using the following method:

1. Use 0.0292 g of NaCl per liter to raise the osmolality 1 mOsmole.
2. Therefore, if you use a 5 M NaCl stock solution (theoretically 10,000 mOsmole), 1 ml/liter will raise the osmolality 10 mOsmole.

Example: Medium A has an osmolarity of 300 mOsmole according to the manufacturer's information. One wishes to study the effect of increasing osmolarity in the range of 300–400 mOsmole.

Experimental Conditions: medium A at 300, 325, 350, 375, and 400 mOsmole.
1. Control 300 mOsmole = Medium A with no additions
2. 325 mOsmole = 10 ml medium A + 25 μl sterile 5 M NaCl solution
3. 350 mOsmole = 10 ml medium A + 50 μl sterile 5 M NaCl solution
4. 375 mOsmole = 10 ml medium A + 75 μl sterile 5 M NaCl solution
5. 400 mOsmole = 10 ml medium A + 100 μl sterile 5 M NaCl solution

Time-Lapse Photomicrography: Assembling Equipment

The following describes the supplies and methods necessary to build a CO_2 controller and heater/temperature controller that can be used to make an incubator around a microscope, such as those shown in Chapter 6. This will work for time-lapse photography, for prolonged viewing of delicate cultures, or for performing experiments that require frequent additions and measurements or photos over the period of a few hours.

Equipment

1. 65 mm Flowmeter, Cole Parmer (1 liter/min), cat# G-03216-70
2. (A) Optronics TEC-470 RGB remote head microscope camera or (B) Cohu 4912 0.5-inch CCD high-resolution monochrome video camera

For CO_2

3. Victor two-stage regulator with needle valve
4. "BioBlend" (5% CO_2, 21% O_2, 74% NO_2) compressed gas, medical grade, size T
5. VCR: Panasonic AG6740 Time Lapse SHVS or capture board (Scion)
6. Software (if capturing to disk): NIH Image (Mac), Scanalytics IPLab (Mac, Windows 95/NT), Improvision OpenLab (Mac)

Materials for Maintaining Temperature and Humidified CO_2 Environment

1. Boxer fan 3.5 inch $\times$ 3.5 inch $\times$ 1.5 inch 9w (Newark Electronics); you can find these locally
2. Aluminum-clad resistors (cat# 13F141 Dale RH50 50W 220 Ω $\pm$1% MC9542) [Newark Electronics (800) 463-9275]

3. Omega 1/16 DIN temperature controller, cat# CN 76133-PV or Whatlow 965
4. Temperature probes: We have used both RTD and thermocouple probes. The thermocouple probes have the advantage of smaller size. We have used both type J (iron/constanton) and type T (copper/constanton). These can be purchased from Omega (cat.# 5TC-TT-T-30-36), Cole Palmer, or Whatlow. RTD probes can be purchased from these same vendors (Whatlow #S00ADT2A036A-X RTD, jacketed leadwire, and thin film RTD, 100 Ω).
5. Screws:
   (3) 6-32$\times \frac{1}{2}$ inch
   (3) 6-32$\times 1\frac{1}{4}$ inch
   (6) 4-40$\times 1$ inch
   Nuts:
   (6) 4–40
6. Miscellaneous:
   (12) Fiber washers for size #6 screw
   (6) fiber spacers $\frac{1}{4}$ inch for size #6 screw
   (3) $1\frac{1}{2}$ inch threaded (6–32) phenolic spacers
   Acrylic: $3\frac{3}{4}$ inch $\times 3\frac{3}{4} \times \frac{5}{16}$ inch
   16–18 g solid or stranded wire
   Heat shrink tubing
7. Acrylic chambers: An acrylic unit, which covers the microscope while allowing access to the microscope stage and controls, can be purchased for most inverted microscope models, or one can be fabricated by a plastics shop. While the small acrylic chamber that sits on the stage is not airtight, the turnover of the 5% CO_2 gas mixture is sufficient to provide adequate CO_2. The gas line runs from the flowmeter to a 1-liter Erlenmeyer flask containing 15 mg/liter phenol red, and from there to a second flask containing copper

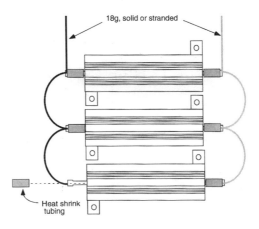

Figure A2.1. Assembly of aluminum-clad resistors.

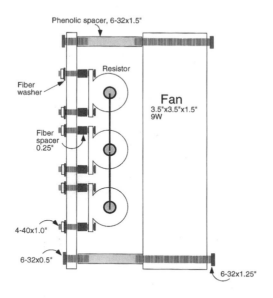

Figure A2.2. Assembly of boxer fan and resistors (side view).

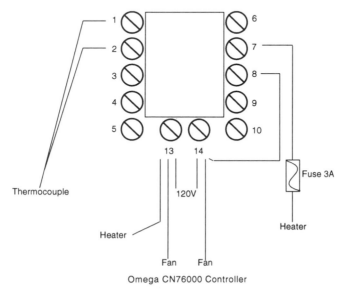

Figure A2.3. Wiring diagram for controller (rear view).

sulfate (4g/liter). The tubing then runs directly to the CO^2 chamber on the stage via a $\frac{1}{4}$-inch hose barb attached to the chamber.

All fittings are John Guest quick disconnect fittings ($\frac{1}{4}$ inch) available from Cole Parmer. All tubing is silicone tubing ($\frac{1}{4}$ inch inner diameter), or, when necessary, less flexible polyethylene tubing that can be used with compression fittings ($\frac{1}{4}$ inch outer diameter).

Procedures

Assembly of the heating system is shown in figures A2.1.–A2.4. The heater unit can be placed to the rear of the microscope stage inside the incubator box. Placing an open dish of sterile water in front of the fan will help humidify the air in dry climates. The flowmeter is used to adjust the flow of the air–CO_2 mixture to compensate for leakage.

We use the CCD camera listed, although other alternatives exist and this technology is improving and becoming cheaper every year. The camera output is sent to a VCR for direct capture on videotape or can be captured to a computer using the software listed. Capture boards and software for image analysis are evolving as well, so it is recommended to contact firmware/software developers through their websites or e-mail to keep current.

A microscope and incubator completely equipped for time-lapse photography are shown in Fig. 6.13.

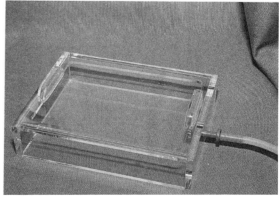

Figure A2.4. Assembled components for heating system, including small acrylic CO_2 chamber.

Pituitary Extract Preparation

Pituitary extract can be prepared in the laboratory. The addition of 2–10 μl/ml of this extract can frequently increase growth of cells in hormone-supplemented, serum-free medium. *Note:* Such medium is not defined, since it is impossible to know what factor in the extract is the active component and different cell types may respond to different components of the pituitary extract.

1. Homogenize 105 g mixed sex bovine pituitaries (Pel Freeze) in 250 ml cold 0.15M NaCl for 10 min in a blender.
2. Transfer the homogenate to a cold beaker and stir for 90 min at 4°C.
3. Centrifuge for 40 min (9800 × G) at 4°C.
4. Discard the pellet.
5. At this point, the supernatant can be aliquoted into 50-ml polypropylene tubes and stored at −20°C.
6. Prior to use, the supernatants should be filter sterilized through a 0.2-μm low-protein-binding filter. *Note:* This can be made easier by centrifuging the thawed supernatants (9800 × G for 20 min at 4°C) to remove particulate material and then passing it through successive 0.8-μm, 0.4-μm, and then 0.2-μm filters.
7. Aliquot into 5-ml snap-cap tubes and store at −20°C.
8. Thawed pituitary extract can usually be stored at 4°C for up to 2 weeks without loss of activity.
9. A dose–response of each batch of extract must be done for each cell line. The extract always has a biphasic dose–response, with too high a concentration being toxic to many cells, especially in serum-free medium.

Siliconization of Glassware

Siliconization of glassware is a method to prevent protein from sticking to the charged walls of glass tubes and other vessels. With the advent of polypropylene tubes and containers with very low protein binding, it is becoming less common to use these procedures. However, sometimes it is still essential to use large glass vessels, such as spinners, in applications where a reduction of protein binding to the vessel is desired. Siliconization of spinners is most essential during suspension adaptation of a cell line to prevent sticking of the cells to the sides of the spinner, with subsequent "caking" of the cells and formation of clumps. Cells also require lower concentrations of some hormones, such as insulin, when grown in serum-free defined medium in siliconized vessels.

The following method works well. Adjust the volume to the size and quantity of glassware to be coated. The coated glassware can be stored for long periods of time (years), and can go through 2–3 wash cycles before recoating is required so it is best to coat enough glassware for required use for several months:

1. Acid wash and dry glass items to be coated.
2. Dip into a 20:1 (v/v) mixture of dimethyldichlorosilane:1,1,1 trichloroethane (the trichloroethane reduces the flashpoint of the mixture)
3. Rinse five times in alcohol.
4. Rinse three times in water.
5. Reassemble and autoclave if sterility is desired.
6. Store dry and sterile for use.

Note: These solutions should be handled with suitable precautions, including using gloves and goggles, working in a well-ventilated area such as a hood, and using proper disposal methods for the chemicals.

Suppliers

The following websites offer the investigator an excellent starting point for locating equipment and supplies:

http://www.faseb.org/bg/
http://www.biosupplynet.com

GENERAL

COLE-PARMER INSTRUMENT CO.
625 E. Bunker Ct.
Vernon Hills, IL 60061
Phone: 847-549-7600
Fax: 847-549-7676
Toll Free: 800-323-4340
E-mail: info@coleparmer.com
Website: www.coleparmer.com/

CURTIN MATHESON SCIENTIFIC
9999 Veterans Memorial Dr.
Houston, TX 77038
Phone: 713-820-9898
Fax: 713-878-3598
Toll Free: 800-323-6428

FISHER SCIENTIFIC CORP.
2000 Park Ln.
Pittsburgh, PA 15275
Phone: 412-490-8300
Fax: 800-926-1166
Toll Free: 800-766-7000

THOMAS SCIENTIFIC
P.O. Box 99
Swedesboro, NJ 08085
Phone: 800-345-2100
Fax: 609-467-3087
E-mail: value@thomassci.com
Website: www.thomassci.com

VWR SCIENTIFIC PRODUCTS
1310 Goshen Parkway
West Chester, PA 19380
Phone: 800-932-5000
Fax: 610-436-1761
Website: www.vwrsp.com

──────── INCUBATORS, CELL CULTURE ────────

FORMA SCIENTIFIC, INC.
P.O. Box 649
Marietta, OH 45750
Phone: 614-373-4763
Fax: 614-373-6770
USA and Canada Toll Free: 1-800-848-3080
E-mail: forma.marketing@lifesciences.com
Website: www.forma.com

BELLCO GLASS INC.
340 Edrudo Road
Vineland, NJ 08360
Phone: 609-691-1075
Fax: 609-691-3247

HERAEUS INSTRUMENTS, INC.
111A Corporate Boulevard
South Plainfield, NJ 07080
Phone: 800-441-2554
Fax: 908-754-9494

HOTPACK CORP.
10940 Dutton Road
Philadelphia, PA 19154
Phone: 215-824-1700
Fax: 215-637-0519
E-mail: hotpack@hotpack.com
Website: www.hotpack.com

LAB-LINE INSTRUMENTS INC.
15th & Bloomingdale Ave.
Melrose Park, IL 60160-1491
Phone: 708-450-2600

Fax: 708-450-0943
Toll Free: 800-LAB-LINE
Website: www.labline.com

QUEUE SYSTEMS
275 Aiken Road
Asheville, NC 28804
Phone: 800-221-4201
Fax: 704-658-0363

THOMAS SCIENTIFIC
P.O. Box 99
Swedesboro, NJ 08085
Phone: 800-345-2100
Fax: 609-467-3087
E-mail: value@thomassci.com
Website: www.thomassci.com

VWR SCIENTIFIC PRODUCTS
1310 Goshen Parkway
West Chester, PA 19380
Phone: 800-932-5000
Fax: 610-436-1761
Website: www.vwrsp.com

TISSUE CULTURE HOODS
(BIOSAFETY CABINETS)

FORMA SCIENTIFIC, INC.
P.O. Box 649
Marietta, OH 45750
Phone: 614-373-4763
Fax: 614-373-6770
USA and Canada Toll Free: 1-800-848-3080
E-mail: forma.marketing@lifesciences.com
Website: http://www.forma.com

HOTPACK CORP.
10940 Dutton Road
Philadelphia, PA 19154
Phone: 215-824-1700
Fax: 215-637-0519
E-mail: hotpack@hotpack.com
Website: www.hotpack.com

INTERMOUNTAIN SCIENTIFIC/BIOEXPRESS
420 N. Kays Dr.
Kaysville, UT 84037
Phone: 801-547-5047
Fax: 801-547-5051

Toll Free: 800-999-2901
E-mail: isc@bioexpress.com
Website: www.bioexpress.com

NUAIRE INCORPORATED
2100 Fernbrook Lane
Plymouth, MN 55447
Phone: 800-328-3352
Fax: 612-553-0459
Website: www.nuaire.com

THE BAKER COMPANY INC.
P.O. Drawer E, Sanford Airport
Sanford, ME 04073
Phone: 207-324-8773
Fax: 207-324-3869
Toll Free: 800-922-2537
E-mail: bakerco@bakerco.com
Website: www.bakerco.com

THOMAS SCIENTIFIC
P.O. Box 99
Swedesboro, NJ 08085
Phone: 800-345-2100
Fax: 609-467-3087
E-mail: value@thomassci.com

VWR SCIENTIFIC PRODUCTS
1310 Goshen Parkway
West Chester, PA 19380
Phone: 800-932-5000
Fax: 610-436-1761

MICROSCOPES

CARL ZEISS, INC.
One Zeiss Drive
Thornwood, NY 10594
Phone: 914-681-7645
Fax: 914-681-7644
Website: www.zeiss.com

LEICA INC.
111 Deer Lake Rd.
Deerfield, IL 60015
Phone: 847-405-0123
Fax: 847-405-0147
Toll Free: 800-248-0123
E-mail: info@leicana.com
Website: www.leica.com

NIKON INC., INSTRUMENT GROUP
1300 Walt Whitman Road
Melville, NY 11747
Phone: 516-547-8500
Fax: 516-547-0306
Website: www.nikonusa.com

OLYMPUS AMERICA INC./P.I.D.
Two Corporate Center Drive
Melville, NY 11747-3157
Phone: 800-446-5967
Fax: 516-844-5112
Website: www.olympus.com

CELL LINES

ATCC (AMERICAN TYPE CULTURE COLLECTION)
12301 Parklawn Dr.
Rockville, MD 20852
Phone: 301-881-2600
Fax: 301-816-4361
Toll Free: 800-638-6597
E-mail: sales@atcc.org
Website: www.atcc.org/

CELL SYSTEMS CORP.
12815 N.E. 124th St., Suite A
Kirkland, WA 98034
Phone: 206-823-1010
Fax: 206-820-6762
Toll Free: 800-697-1211

CLONETICS CORP.
9620 Chesapeake Drive
Suite 201
San Diego, CA 92123
Phone: 619-541-0086
Fax: 619-541-0823
Website: www.clonetics.com

CHEMICALS AND OTHER REAGENTS

(ALAMAR BLUE)
ACCUMED INTERNATIONAL, INC.
900 N. Franklin, Suite 401
Chicago, IL 60610
Phone: 800-650-2228
E-mail: info@accumed.com

BOEHRINGER MANNHEIM CORP.
Boehringer Mannheim Biochemicals Div.
9115 Hague Rd.
P.O. Box 50414
Indianapolis, IN 46250
Phone: 800-428-5433
Fax: 317-576-7317
Toll Free: 800-262-1640
E-mail: biochemts-us@bmc.boehringer-mannheim.com
Website: www.biochem.boehringer-mannheim.com

J.T. BAKER
Mallinckrodt-Baker Inc. Div.
222 Red School Ln.
Phillipsburg, NJ 08865
Phone: 908-859-2151; 800-582-2637
Fax: 908-859-9318

MOLECULAR PROBES, INC.
4849 Pitchford Avenue
P.O. Box 22010
Eugene, OR 97402-9144
Phone: 800-438-2209
Fax: 541-334-6504
Website: www.probes.com

RESEARCH ORGANICS, INC.
4353 East 49th Street
Cleveland, OH 44125
Phone: 800-321-0570
Fax: 216-883-1576
E-mail: info@resorg.com
Website: www.resorg.com

PHARMACIA BIOTECH INC.
800 Centennial Avenue
P.O. Box 1327
Piscataway, NJ 08855-1327
Phone: 800-526-3593
Fax: 800-FAX-3593
Website: www.biotech.pharmacia.se

SIGMA CHEMICAL CO.
P.O. Box 14508
St. Louis, MO 63178
Phone: 800-521-8956
Fax: 800-325-5052
E-mail: custserv@sial.com
Website: www.sigma.sial.com

WORTHINGTON BIOCHEMICAL CORP.
450 Halls Mill Road

Freehold, NJ 07728
Phone: 800-445-9603
Fax: 800-368-3108

─────────── **GROWTH FACTORS** ───────────

BECTON DICKINSON-LABWARE
2 Oak Park
Bedford, MA 01730
Phone: 617-275-0004
Fax: 617-275-0043
Toll Free: 800-343-2035
E-mail: mail@cbpi.com
Website: www.cbpi.com

BIOMEDICAL TECHNOLOGIES, INC.
378 Page Street
Stoughton, MA 02072
Phone: 617-344-9942
Fax: 617-341-1451

BIOWHITTAKER, INC.
8830 Biggs Ford Road
Walkersville, MD 21793-0127
Phone: 301-898-7025
Fax: 301-845-8338
E-mail: sales@biowhittaker.com
Website: www.biowhittaker.com

BOEHRINGER MANNHEIM CORP.
Boehringer Mannheim Biochemicals Div.
9115 Hague Rd.
P.O. Box 50414
Indianapolis, IN 46250
Phone: 800-428-5433
Fax: 317-576-7317
Toll Free: 800-262-1640
E-mail: biochemts-us@bmc.boehringer-mannheim.com
Website: www.biochem.boehringer-mannheim.com

CALBIOCHEM-NOVABIOCHEM CORP.
10394 Pacific Ctr. Ct.
San Diego, CA 92121
Phone: 619-450-9600
Fax: 800-776-0999
Toll Free: 800-854-3417
E-mail: orders@calbiochem.com
Website: www.calbiochem.com

CLONETICS CORP.
9620 Chesapeake Drive

Suite 201
San Diego, CA 92123
Phone: 619-541-0086
Fax: 619-541-0823
Website: www.clonetics.com

LIFE TECHNOLOGIES, INC.
P.O. Box 6009
Gaithersburg, MD 20884
Phone: 301-840-4000
Fax: 301-670-8539
Website: www.lifetech.com

R&D SYSTEMS
614 McKinley Place NE
Minneapolis, MN 55413
Phone: 612-379-2956
Website: www.rndsystems.com

SIGMA CHEMICAL CO.
P.O. Box 14508
St. Louis, MO 63178
Phone: 800-521-8956
Fax: 800-325-5052
E-mail: custserv@sial.com
Website: www.sigma.sial.com

UPSTATE BIOTECHNOLOGY
199 Saranac Ave.
Lake Placid, NY 12946
Phone: 518-523-1518
Fax: 781-890-7738
Toll Free: 800-233-3991
E-mail: info@upstatebiotech.com
Website: www.upstatebiotech.com

IMMUNOHISTOCHEMISTRY

BECTON DICKINSON IMMUNOCYTOMETRY SYSTEMS
2530 Qume Drive
San Jose, CA 95131
Phone: 408-954-2128
Fax: 408-954-2009
Website: www.cbpi.com

BIODESIGN INTERNATIONAL
105 York Street
Kennebunk, ME 04043
Phone: 207-985-1944

Fax: 207-985-6322
Website: www.biodesign.com

BIOMEDICAL TECHNOLOGIES, INC.
378 Page Street
Stoughton, MA 02072
Phone: 617-344-9942
Fax: 617-341-1451

CHEMICON INTERNATIONAL, INCORPORATED
28835 Single Oak Drive
Temecula, CA 92590
Phone: 800-437-7500
Fax: 909-676-9209
Website: www.chemicon.com

JACKSON IMMUNORESEARCH LABORATORIES, INC.
872 West Baltimore Pike
West Grove, PA 19390
Phone: 800-367-5296
Fax: 610-869-0171
E-mail: cuserjaxn@aol.com

MOLECULAR PROBES, INC.
4849 Pitchford Avenue
P.O. Box 22010
Eugene, OR 97402-9144
Phone: 800-438-2209
Fax: 541-334-6504
Website: www.probes.com

PENINSULA LABORATORIES, INC.
611 Taylor Way
Belmont, CA 94002
Phone: 800-922-1516
Fax: 415-592-5392
Website: www.penlabs.com

R&D SYSTEMS
614 McKinley Place NE
Minneapolis, MN 55413
Phone: 612-379-2956
Website: www.rndsystems.com

SANTA CRUZ BIOTECHNOLOGY, INC.
2161 Delaware Avenue
Santa Cruz, CA 95060
Phone: 800-457-3801
Fax: 408-457-3801
E-mail: scbt@netcom.com
Website: www.scbt.com

VECTOR LABORATORIES INC.
30 Ingold Road
Burlingame, CA 94010
Phone: 415-697-3600
Fax: 415-697-0339

ZYMED LABORATORIES, INC.
458 Carlton Court
South San Francisco, CA 94080
Phone: 415-871-4494
Fax: 415-871-4499
E-mail: tech@zymed.com

IMAGING, ANALOG, DIGITAL, AND SOFTWARE

COMPIX INC., IMAGING SYSTEMS (DOS, Windows 95)
705 Thomson Park Drive
Cranberry Township, PA 16066-6426
Phone: 412-772-5277
Fax: 412-772-5278
Website: www.compix.com

DAGE-MTI, INC.
701 N. Roeske Avenue
Michigan City, IN 46360
Phone: 219-872-5514
Fax: 219-872-5559
E-mail: dage@adsnet.com

FOTODYNE INC.
950 Walnut Ridge Drive
Hartland, WI 53029
Phone: 800-362-3686
Fax: 414-369-7017
E-mail: fotodyne@aol.com

HAMAMATSU PHOTONIC SYSTEMS CORP.
360 Foothill Road
Bridgewater, NJ 08807-0910
Phone: 908-231-1116
Fax: 908-231-0852
Website: www.hamamatsu.com

IMPROVISION INC. USA (Mac OS)
Suite 370
95 Old Colony Avenue
Boston, MA 02170
Phone: 617-745-0002; 888-Openlab
Fax: 617-745-0003

E-mail: admin@improv.co.uk
Website: www.improvision.com

OPTRONICS ENGINEERING
175 Cremona Dr.
Goleta, CA 93117
Phone: 805-968-3568
Fax: 805-968-0933
Website: www.optronics.com

PHOTOMETRICS LTD.
3440 E. Britannia Drive
Tucson, AZ 85706
Phone: 520-889-9933
Fax: 520-573-1944
Website: www.photomet.com

SIGNAL ANALYTICS CORP. (Mac OS, Windows 95/NT)
4440 Maple Avenue East
Suite 201
Vienna, VA 22180
Phone: 703-281-3277
Fax: 703-281-2509
E-mail: sdowds@iplab.com
Website: www.iplab.com

NIH IMAGE HOME PAGE (Mac OS)
http://rsb.info.nih.gov/NIH-Image/Default.html

PLATE READERS

MOLECULAR DEVICES
1311 Orleans Drive
Sunnyvale, CA 94089
Phone: 408-747-3559
Fax: 408-747-3602
Website: www.moldev.com

MOLECULAR DYNAMICS
928 East Arques Avenue
Sunnyvale, CA 94086
Phone: 800-333-5703
Fax: 408-773-1493
E-mail: info@mdyn.com
Website: www.mdyn.com

PERSEPTIVE BIOSYTEMS
500 Old Connecticut Path
Framingham, MA 01701
Phone: 800-899-5858

Fax: 508-388-7880
Website: www.pbio.com

SLT LABINSTRUMENTS (Tecan, USA)
P.O. Box 13953
Research Triangle Park, NC 27709
Phone: 919-361-5200
Fax: 919-361-5201

EG&G WALLAC
Analytical Systems Div.
9238 Gaither Rd.
Gaithersburg, MD 20877
Phone: 301-963-3200
Fax: 301-963-7780
Toll Free: 800-638-6692
E-mail: 104102.1737@compuserve.com
Website: www.wallac.com

ANIMAL SUPPLIERS

CHARLES RIVER LABORATORIES
251 Ballardvale St.
Wilmington, MA 01887
Phone: 978-658-6000
Fax: 978-658-7132
Toll Free: 800-LAB-RATS
E-mail: comments@criver.com
Website: www.criver.com

HARLAN SPRAGUE DAWLEY, INC.
P.O. Box 29176
Indianapolis, IN 46229
Phone: 317-899-7511
Fax: 317-899-1766

HILLTOP LAB ANIMALS INC.
Hilltop Drive
Scottsdale, PA 15683
Phone: 800-245-6291
Fax: 412-887-3582
E-mail: edmied@aol.com

CRYOGENIC EQUIPMENT

BARNSTEAD/THERMOLYNE CORP.
2555 Kerper Boulevard
Dubuque, IA 52001

Phone: 319-556-2241
Fax: 319-556-0695
Website: www.barnsteadthermolyne.com

REVCO SCIENTIFIC
One Revco Drive
Asheville, NC 28804
Phone: 800-252-7100
Fax: 704-645-3368

TAYLOR-WHARTON CRYOGENICS
A Div. of Harsco Corp.
P.O. Box 8316
Camp Hill, PA 17001-8316
Phone: 717-763-5060
Fax: 717-763-5061
Toll Free: 800-898-2657
Website: www.taylor-wharton.com

BALANCES, pH METERS

COLE-PARMER INSTRUMENT CO.
625 E. Bunker Ct.
Vernon Hills, IL 60061
Phone: 847-549-7600
Fax: 847-549-7676
Toll Free: 800-323-4340
E-mail: info@coleparmer.com
Website: www.coleparmer.com

INTERMOUNTAIN SCIENTIFIC/BIOEXPRESS
420 N. Kays Dr.
Kaysville, UT 84037
Phone: 801-547-5047
Fax: 801-547-5051
Toll Free: 800-999-2901
E-mail: isc@bioexpress.com
Website: www.bioexpress.com

OHAUS CORPORATION
29 Hanover Road
Florham Park, NJ 07932
Phone: 201-377-9000
Fax: 201-593-0359
Website: www.ohaus.com

ORION RESEARCH, INC.
500 Cummings Center
Beverly, MA 01915

Phone: 800-225-1480
Fax: 508-927-3932
E-mail: webmaster@orionres.com
Website: www.orionres.com

RADIOMETER ANALYTICAL PRODUCTS GROUP
810 Sharon Drive
West Lake, OH 44145
Phone: 1-440-871-5975
Fax: 1-440-899-1139

SARTORIUS CORP.
131 Heartland Boulevard
Edgewood, NY 11717
Phone: 516-254-4249
Fax: 516-254-4253
E-mail: 102233.432@compuserve.com
Website: www.sartorius.com

THOMAS SCIENTIFIC
P.O. Box 99
Swedesboro, NJ 08085
Phone: 800-345-2100
Fax: 609-467-3087
E-mail: value@thomassci.com

VWR SCIENTIFIC PRODUCTS
1310 Goshen Parkway
West Chester, PA 19380
Phone: 800-932-5000
Fax: 610-436-1761

MISCELLANEOUS TISSUE CULTURE SUPPLIES AND EQUIPMENT

BELLCO GLASS INC. (Spinner bottles, cloning rings)
340 Edrudo Road
Vineland, NJ 08360
Phone: 609-691-1075
Fax: 609-691-3247

BECKMAN INSTRUMENTS, INC. (centrifuges)
2500 Harbor Boulevard
Fullerton, CA 92634
Phone: 800-742-2345
Fax: 800-643-4366
Website: www.beckman.com

BRINKMANN INSTRUMENTS, INC. (liquid handling)
One Cantiague Road

P.O. Box 1019
Westbury, NY 11590-0207
Phone: 516-334-7500
Fax: 516-334-7506
E-mail: info@brinkmann.com
Website: www.brinkmann.com

CORNING LABWARE & EQUIPMENT (general lab supplies, tissue culture
plasticware)
Science Products Div.
P.O. Box 5000
Corning, NY 14830
Phone: 607-974-7740
Fax: 607-974-0345
Toll Free: 800-222-7740

EPPENDORF SCIENTIFIC, INC. (sample preparation, liquid handling, centrifuges)
6524 Seybold Road
Madison, WI 53719
Phone: 608-276-9855
Website: www.eppendorf.com

FINE SCIENCE TOOLS INC. (microdissection supplies)
6 Ohio Drive
373-G Vintage Park Drive
Foster City, CA 94404
Phone: 800-521-2109
Fax: 415-349-3729
E-mail: info@finescience.com
Website: www.finescience.com

MILLIPORE CORP. (water purification equipment, filtration)
80 Ashby Road
Bedford, MA 01730
Phone: 800-645-5476
Fax: 617-275-5550
Website: www.millipore.com

MILTEX INSTRUMENT COMPANY (microdissection supplies)
6 Ohio Drive
Lake Success, NY 10042
Phone: 516-775-7100, 800-645-8000
Fax: 516-775-7185

NALGE NUNC INTERNATIONAL (tissue culture plasticware)
2000 North Aurora Road
Naperville, IL 60563
Phone: 708-416-2122
Fax: 708-416-2556

OMEGA ENGINEERING INC. (temperature controllers)
One Omega Drive, P.O. Box 4047
Stamford, CT 06907-0047

Phone: 203-359-1660
Fax: 203-359-7700
E-mail: info@omega
Website: www.omega.com

PEL-FREEZ BIOLOGICALS (animal tissue supplier)
205 N. Arkansas Street
Rogers, AR 72757
Phone: 800-643-3426
Fax: 501-636-3562
Website: www.pelfreez-bio.com

RAININ INSTRUMENT CO. (pipetting equipment)
Mack Road, Box 4026
Woburn, MA 01888-4026
Phone: 617-935-3050
Fax: 617-938-1152
Website: www.rainin.com

ROBOZ SURGICAL INSTRUMENT CO. (microdissection supplies)
9210 Corporate Boulevard
Suite 220
Rockville, MD 20850
Phone: 301-590-0055
Fax: 301-590-1290

SOLOHILL ENGINEERING INC. (microcarriers)
4220 Varsity Dr.
Ann Arbor, MI 48105
Phone: 313-973-2956
Fax: 313-973-3029

WHATLOW (temperature controllers)
12001 Lackland Road
St. Louis, MO 63146
Phone: 314-878-4600
Fax: 314-878-6814

Index